Metric Graphs for Interpretation

METRIC GRAPHS
FOR INTERPRETATION

GORDON L. BELL M.A.

*Principal Teacher of Mathematics
Forrester Secondary School
Edinburgh*

GEORGE G. HARRAP & CO. LTD
London Toronto Wellington Sydney

First published in Great Britain 1972
by GEORGE G. HARRAP & CO. LTD
182–184 High Holborn, London WC1V 7AX

© *Gordon L. Bell* 1972

All rights reserved. No part of this publication may be reproduced in any form or by any means without the prior permission of George G. Harrap & Co. Ltd

ISBN 0 245 50565 2

Text set in 11/12 pt Monotype Baskerville, printed by letterpress, and bound in Great Britain at The Pitman Press, Bath

PREFACE

The Aims of this book are:
1. To interest and train pupils in interpreting graphs by providing many of these graphs already drawn.
2. To provide a natural sequence which will simplify progress from type to type.
3. To introduce at appropriate stages such essential features as scales, variables, continuity, etc.
4. To give the pupils just sufficient practice in drawing their own graphs to ensure that they can apply the knowledge gained.

As the title indicates, metric measurements are used throughout the book which lays special emphasis on graphs suitable for pupils who are preparing for
   (a) CSE examinations in Mathematics, or
   (b) GCE or SCE examinations in Arithmetic.

The inclusion of 'Time Series and Trend' graphs (Chapter 9) and Statistical graphs (Chapter 10) should prove of additional value to pupils who intend to enter business or commerce.

G.L.B.

CONTENTS

CHAPTER		PAGE
I.	PICTOGRAPHS	9
II.	PIE CHARTS	18
III.	COLUMN GRAPHS	23
IV.	LINE GRAPHS	35
V.	JAGGED LINE GRAPHS	41
VI.	CURVED GRAPHS	57
VII.	STRAIGHT LINE GRAPHS	71
VIII.	SPEED, TIME AND DISTANCE GRAPHS	87
IX.	TIME SERIES AND TREND GRAPHS	105
X.	STATISTICAL GRAPHS	131

Watching TV

 Represents 1 viewer

Chapter I

PICTOGRAPHS

In newspapers and magazines you will often be attracted by rows of little men, motor cars, or houses. This is the editor's way of helping you to understand important facts which you would pass by if they were given in the form of numbers. They are used to catch the eye, to make a quick impression and especially if the information they give does not need to be very exact.

Example 1

Watching TV

The pictograph on the opposite page shows how many hours per week, on the average, were spent watching television by various members of a school class of 36.

Study the seated figure on the top line which gives the scale of the graph, and then answer the questions in the exercises.

Exercises

(1) The scale is: "1 seated figure represents . . . viewer."

(2) Say how many pupils watched for the following number of hours:

(*a*) 1 to 4, (*b*) 9 to 12, (*c*) 17 to 20.

(3) State for which numbers of hours the following number of pupils watched TV:

(*a*) 4, (*b*) 12, (*c*) 2.

(4) Say what you understand by the figure in the last row, sitting with his head bowed.

(5) If any time over 16 hours per week is likely to retard a pupil's progress at school, what fraction of this class would be so affected?

(6) Say how many pupils watch TV:

(*a*) for 13 hours or more per week,
(*b*) for 8 hours or less per week.

Completed Houses

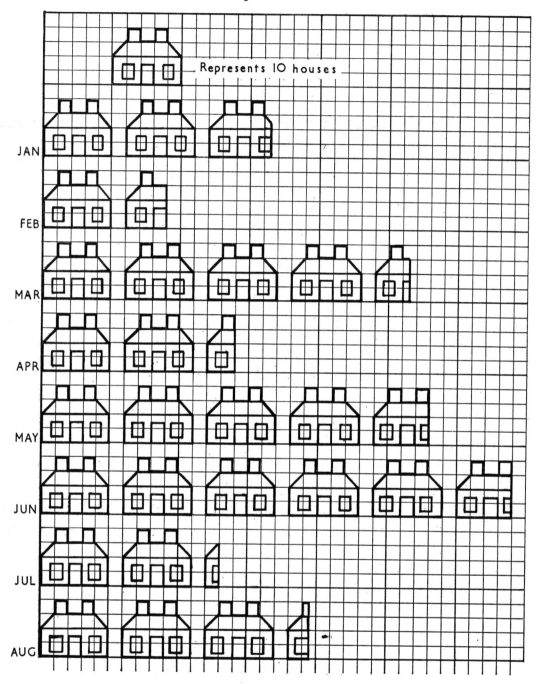

Example 2

Completed Houses

This pictograph shows how many houses were finished in a town during 8 months of a year.

When you draw a pictograph try to make the drawing simple like this one, and the same on either side of a central line (symmetrical).

Notice the scale, or what 1 house stands for, and also how, as a result of symmetry, a part picture can be made to represent an exact number of houses. Yet the scale is *smaller* than in Example 1, because 1 actual house is represented by only $\frac{1}{10}$ of the scale house in the pictograph.

Exercises

(1) How many divisions wide is each house in the pictograph?

(2) State how many actual houses are shown by a part picture of this width:

  (*a*) $2\frac{1}{2}$ divisions, (*b*) 2 divisions, (*c*) 3 divisions.

(3) State how many divisions wide a part picture will be to represent:

  (*a*) 8 houses, (*b*) 3 houses.

(4) State how many houses were completed during:

  (*a*) January, (*b*) March, (*c*) July.

(5) Say in which months this number of houses was completed:

  (*a*) 16, (*b*) 24, (*c*) 33.

(6) Say during which month the number of houses completed was:

  (*a*) half the largest number,

  (*b*) three times the number for February.

(7) Bad weather affected certain months. Which two months were worst hit?

Number of Cars Crossing a Toll Bridge

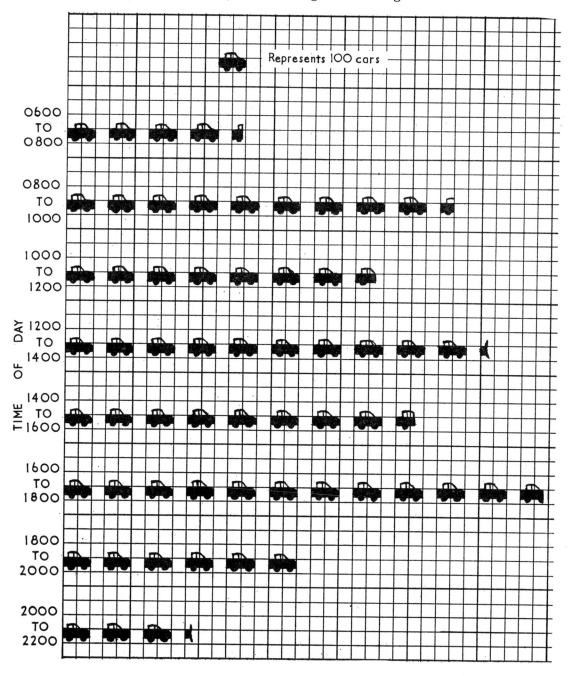

Example 3

Number of Cars Crossing a Toll Bridge

The pictograph showing the flow of traffic over a toll bridge on a normal weekday requires the use of a small scale. This means that each actual car is shown by a very small fraction of the car in the pictograph. It is not possible to read the graph accurately, but you must study the scale and try to judge the numbers of cars to the nearest ten when answering the exercises.

Exercises

(1) Say how many cars are represented by a part 'pictocar' of width:

(*a*) 1 division, (*b*) 0·2 of 1 division, (*c*) 0·6 of 1 division, (*d*) 1·4 divisions.

(2) State during which periods of time the following number of cars crossed the toll bridge:

(*a*) 770, (*b*) 430, (*c*) 1180.

(3) Say how many cars crossed the toll bridge during the periods:

(*a*) 0800–1000 hours, (*b*) 1400–1600 hours, (*c*) 1800–2000 hours.

(4) Say during which period the number of cars crossing the bridge was:

(*a*) twice the number for 0600–0800 hours,

(*b*) approximately three times the number for 2000–2200 hours.

(5) (*a*) State the three periods during which most cars crossed the bridge.

(*b*) Can you suggest why the flow of traffic was greatest during these times?

(6) Find

(*a*) the total number of cars crossing the bridge between 0600 hours and 2200 hours, and

(*b*) the average number of cars crossing the bridge *per hour* to the nearest ten cars.

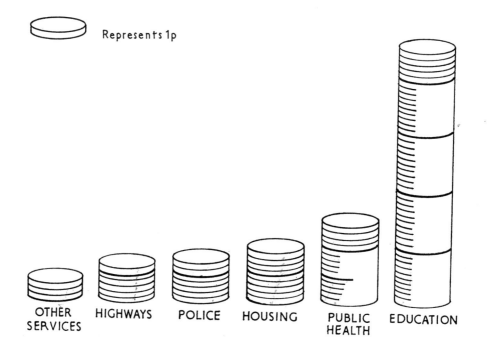

Example 4

City Spending

This pictograph shows how a large city in Britain spent every £1 it collected in rates in a recent year. Notice how each penny (p) is shown in the drawing. The scale for this pictograph is a **large one,** and allows us to read exact amounts from the graph.

Exercises

(1) State how much of each £1 collected is spent on

(*a*) children, (*b*) police, (*c*) public health.

(2) Say on which services the following amounts in each £1 are spent:

(*a*) 4p, (*b*) 6p, (*c*) **45p.**

(3) State on how many items there was

(*a*) less than 5p in every £ spent,
(*b*) between 5p and 15p in every £ spent,
(*c*) more than 15p in every £ spent.

(4) Express the amount spent on

(*a*) education, (*b*) public health and (*c*) highways

(i) as a decimal part of £1,
(ii) as a percentage of £1.

(5) If the actual amount from the rates allocated to education was **£12 312 000,** find

(*a*) the total raised in rates,
(*b*) the amount raised by each 1p of the rate levied ('the product of a penny rate').

Note: Notice how the pennies shown piled one on top of the other form columns. Could the heights of these columns tell the same story as our pictograph?

Look back at your pictographs 1, 2 and 3. Turn each one sideways and see if you can spot the same type of columns.

EXERCISES ON DRAWING PICTOGRAPHS

(1) The number of teachers in the various departments of a comprehensive school are: English 8, Mathematics 7, Modern studies 4, Technical 5, Science 9, Physical education 4, Art 5, Music 3.

Illustrate this by means of a pictograph, using a single unit scale (1 figure represents 1 person). ♟ You may draw your figures like this one equally spaced out on blank paper.

(2) The number of goals scored in a season by a school's six leading goal scorers was:

   Hall 32,   Graham 27, Turnbull 25,
   Morgan 20, Smith 18,   Welsh 17.

Take as your scale picture a circle divided into 4 quadrants: to represent 4 goals. Decide how you would show (say) 3 goals, rule equally spaced lines on blank paper and space your circles evenly along these lines. The top line is for the scale circle.

(3) The number of bottles of free milk supplied per day to a primary school between 1945 and 1970 is shown at intervals of 5 years:

   1945 240   1950 340   1955 470
   1960 630   1965 600   1970 510

Use 0·5 cm squared paper and let your scale bottle (drawn as in the sketch) represent 40 bottles of milk.

(4) The number of Christmas trees sold by a multiple store in the 7 weeks before Christmas was:

Week	1	2	3	4	5	6	7
Number of trees sold	24	30	42	66	100	126	82

When you draw your trees you may change the numbers in the table to dozens to help you choose a suitable scale.

Chapter **II**

PIE CHARTS

Pie charts (or circle diagrams) are used to show the various parts into which a whole is divided. As the name suggests, each chart may be thought of as a circular pie divided into slices called sectors. The fraction of the whole allotted to each item is equal to the fraction of 360° which the corresponding sector subtends at the centre of the circle.

Examples of information best shown on a pie chart are:

(1) the allocation of periods in a school week to various subjects,
(2) the division of each £1 of rates among various items of city expenditure,
(3) the allocation of the family income to food, fuel, rent, etc.

Since in many cases we divide the whole (= 100 per cent) into parts and then have to express each part as a fraction of 360° (which corresponds to 100 per cent), it may be useful to have a ready reckoner in the form of protractor (see Fig. 1 below). This can be traced and the tracing used for drawing the pie chart when the various percentages are known.

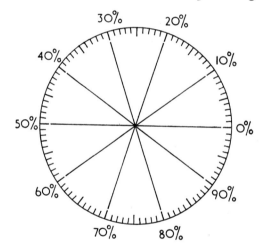

Fig. 1

Example 1

Fig. 2 shows the allocation of time per week to the various subjects for pupils following a language course. This covers 40 school periods per week.

The language sector subtends a 90° angle at the centre of the circle and as

$$\frac{90°}{360°} = \tfrac{1}{4},$$

the number of language periods per week is $\tfrac{1}{4}$ of 40 which equals 10.

Exercises

(1) Measure each of the other sector angles and complete a table giving the number of periods per week allocated to each subject, using the following headings:

(*a*) Subject, (*b*) Sector angle, (*c*) Number of periods per week.

(2) Suggest how you would apportion the time for pupils following a science course.

Time-Table for Language Course   Consumer Expenditure

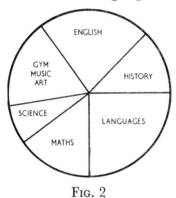

Fig. 2   Fig. 3

Example 2

Fig. 3, which represents the total consumer expenditure of £3,930 million in Great Britain during a recent quarter of a year, shows how this was divided among the main items.

Exercises

(1) Measure the sector angles and compile a table to show:

(*a*) Item(s), (*b*) Sector angle.

(2) Find from your table, correct to 1 decimal place, the percentage of the total expenditure devoted to:

(*a*) Housing, fuel and light, (*b*) Household goods, (*c*) Cars.

(3) Find also the total value of expenditure during the quarter, to the nearest £million, on:

(*a*) Clothing, (*b*) Food, drink and tobacco.

Example 3

Figs. 4 and 5 show the division of Mr Dow's income per month during the years 1970 and 1971. The areas of the circles are in proportion to his monthly income for these years. In 1970 Mr Dow was a Sales Representative earning £100 per month, and in 1971 he was Sales Manager.

Exercises

(1) Measure the radii of the two circles and calculate the ratio of their areas.
(2) Hence calculate Mr Dow's income per month in 1971.
(3) Compare the monthly cash amounts allocated to clothes in 1970 and in 1971.
(4) Say which items have equal shares of the income: (*a*) in 1970 and (*b*) in 1971.
(5) (*a*) Which items represent a smaller percentage of the monthly income in 1971 than in 1970?
   (*b*) By comparing the cash amounts for these items, show that they are all greater in 1971 than in 1970.
(6) Calculate the sector angle for housekeeping in: (*a*) 1970 and (*b*) 1971.
(7) If one-third of the increase in expenditure on rent and rates is caused by increased rating, how much more does Mr Dow pay for rates in 1971?

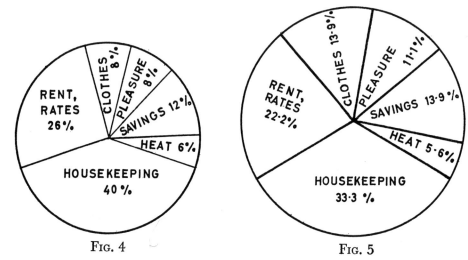

Fig. 4   Fig. 5

Note: It will now be realized that pie charts have limited value owing to:
   (1) the difficulty of gauging and comparing the sizes of the sectors without calculation, measurement, or percentage figures in the sectors;
   (2) the lack of accuracy.

EXERCISES ON CHAPTER II

(1) Each £1 of sales revenue of a firm is accounted for as follows: wages 35p; materials 21p; expenses 12½p; taxation 22½p; profit 9p. Using a circle of radius 5 cm, draw a pie chart to represent this division.

(2) A Life Assurance Society invested each £1 of premiums paid by policyholders as follows:

Ordinary shares	42½p
Government securities	25p
Property shares	10p
Debentures	15p
Local loans, etc.	7½p

(a) Think of these amounts in the £ (100p) as percentages.
(b) Convert these to sector angles and illustrate the division on a pie chart of radius 4 cm.

(3) In 1969 the supply of electricity to consumers in a certain area was divided as follows: domestic 30%; commercial 15%; industrial 51%; farm and public lighting 4%. Draw a circle of radius 4 cm and, by *calculating* the necessary sector angles, draw a circle diagram to illustrate these percentages.

(4) In Scotland in 1967 the principal classes of consumers and the corresponding percentages of total gas supplied were:

Consumer	Percentage
Domestic prepayment	40·3
Domestic credit	18·9
Industrial	21
Commercial	14·3
Public administration, lighting, etc.	5·5

(a) Show these divisions on a pie chart of radius 4 cm.
(b) Illustrate the same figures by a bar graph.
(c) Discuss the merits of the two graphs with regard to:
   (i) Ease of drawing.
   (ii) Presentation of facts individually.
   (iii) Presentation of facts as part of a whole.
   (iv) Accuracy.

(5) The pictograph (Fig. 6) is used to illustrate that the ratio of sheet steel production for a year in three areas of the United Kingdom was 4·5:3:1.
   (a) Check that each of the dimensions of length, breadth and thickness measured in millimetres is approximately in this ratio.
   (b) State, with reasons, why you consider that the pictograph does not give a clear picture of this relationship.
   (c) Using the diagram for Area C as 1 unit, draw a *sketch* which gives a more accurate comparison of production in the three areas.

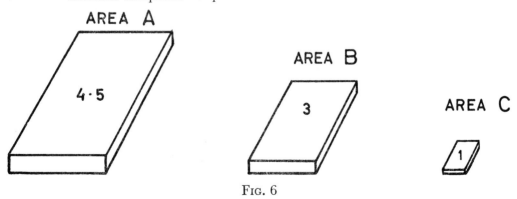

Fig. 6

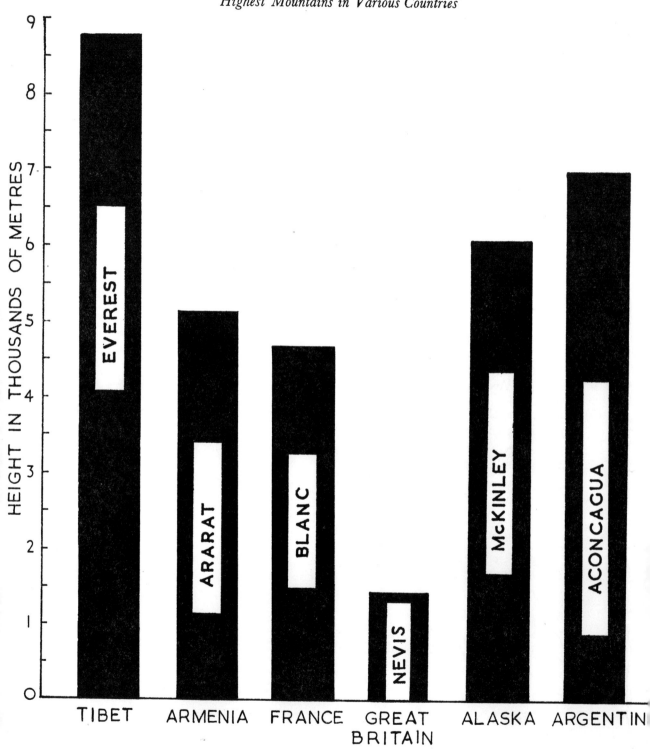

Chapter III

COLUMN GRAPHS

While the pictographs we have studied form an attractive way of presenting facts and changes they have two main drawbacks:

(1) They take a long time to draw.
(2) They are difficult to read accurately.

Yet they point the way, as you saw, to showing the same types of facts by means of columns called *Column Graphs* or *Bar Graphs*. Column Graphs are more easily drawn than Pictographs, and can be read more accurately.

When you have studied the column graphs which follow you will see how clearly they show changes, and how they help you to compare quantities singly or in groups.

Example 1

Highest Mountains in Various Countries

Note first that the height is shown in thousands of metres, so that 1 stands for 1000, 2 for 2000, and so on. Now place your ruler along this 'Height' line and check that it is marked off in centimetres.

Exercises

(1) What is the scale for height? Decide not only what 1 cm represents, but also what 0·1 cm represents.
(2) Say what height would be shown by columns
    (*a*) 5 cm high.
    (*b*) 8·2 cm high.
    (*c*) 14·7 cm high.
(3) Say what height of column would show mountains
    (*a*) 3000 metres high.
    (*b*) 5500 metres high.
    (*c*) 7600 metres high.
(4) The highest mountain shown is . . . in . . .
(5) The smallest mountain shown is . . . in . . .
(6) (*a*) Which two mountains are nearly equal in height?
    (*b*) Guess the difference in height between these two mountains.
(7) How many times as high as 'NEVIS' is 'BLANC'?
(8) Measure each mountain in centimetres and tenths of a centimetre, make out a table, and find the actual heights, using the scale you found in answer to Exercise 1.

Mountain	Column height	Actual height
Everest	17·6 cm	

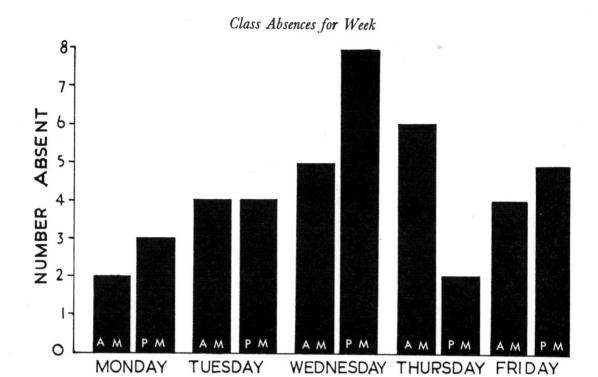

Class Absences for Week

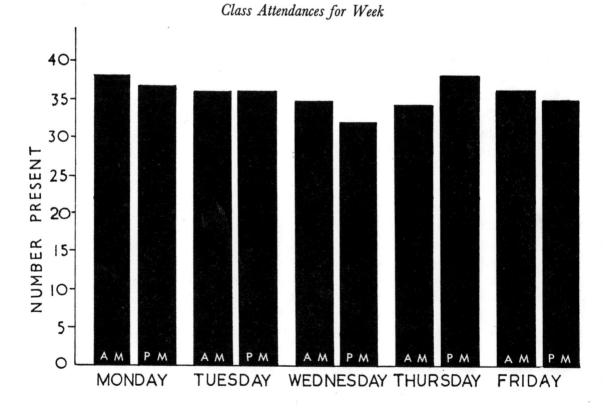

Class Attendances for Week

Example 2

Class Absences for Week

You will see that the height of each column shows how many pupils were absent when the class register was marked.

Exercises

(1) How many times per week was the register marked?

(2) What is the scale for the number absent? (Use the cm scale on your ruler if necessary.)

(3) Say how many pupils were absent on

(*a*) Monday A.M., (*b*) Wednesday P.M., (*c*) Friday P.M.

(4) Find when the following numbers of pupils were absent:

(*a*) 3, (*b*) 5, (*c*) 6.

(5) Find on which day

(*a*) the number of absentees increased most in the afternoon,

(*b*) the number of absentees decreased most in the afternoon.

(6) Which day showed the same number of absences in the afternoon as in the morning?

(7) Say which afternoon showed the same number of absences as

(*a*) Monday morning, (*b*) Friday morning.

Example 3

Class Attendance for Week

You will require a ruler again to answer most of these questions.

Exercises

(1) What do the heights of columns tell us in this graph?

(2) What is the scale for this quantity? (You must decide what 1 cm represents, and how many mm represent 1 pupil attendance.)

(3) Find how many pupils were present

(*a*) on Monday A.M., (*b*) on Wednesday P.M., (*c*) on Friday P.M.

(4) Say when the following numbers of pupils were present:

(*a*) 37, (*b*) 35, (*c*) 34.

(5) Find on which day

(*a*) the number present dropped most in the afternoon,

(*b*) the number present rose most in the afternoon.

(6) As these two graphs refer to the same week for the same class, how many pupils were in the class?

Compare the two graphs Examples 2 and 3, which give the same information in a different way, and you will notice that:

(i) On the upper graph each pupil is shown by a column 1 cm high (larger scale), whereas on the lower graph each pupil present is shown by a column 0·2 cm high (smaller scale).

(ii) Any change is made much more obvious by the rise or fall in column height in the absence graph than in the attendance graph.

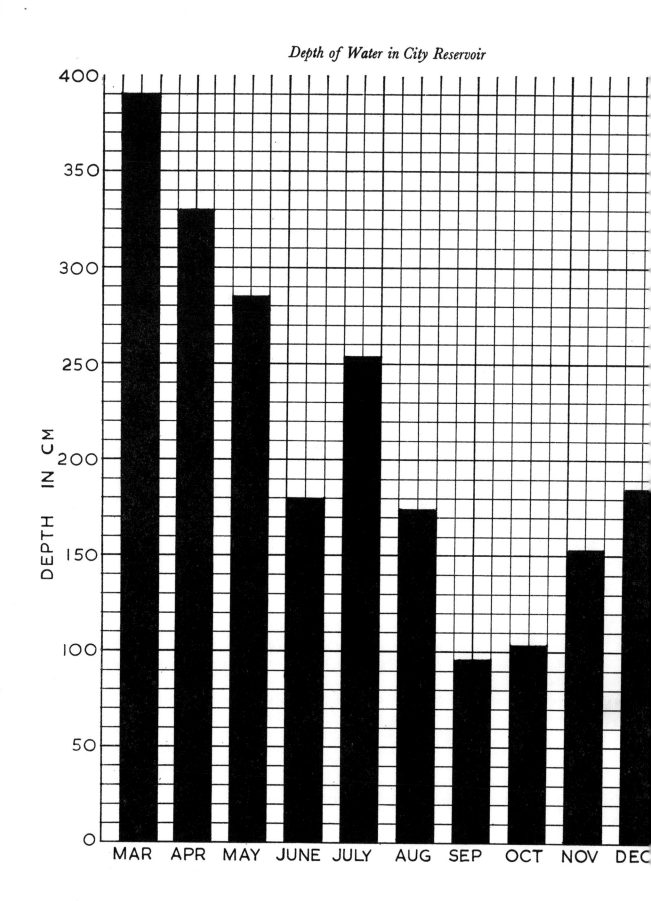

Example 4

Depth of Water in a City Reservoir

You should study the title of a graph as well as what the columns show. Here you can imagine that a long dip-stick has been used to test the depth of the water, and that the columns show the wet part. The graph shows the depth of water in a city reservoir on the last day of each month from March to December of a year.

If Column Graphs are drawn on squared paper, like the one opposite, they can be read more easily and more accurately. The 2·5 cm squares have been divided into 0·5 cm squares so that you need not use a ruler. See how easily you can answer the following questions.

Exercises

(1) The scale for depth is

    1 large division (2·5 cm) =   cm

    2 small divisions (1 cm) =   cm

(2) Use this scale to fill in the blanks in the following list:

	(a)	(b)	(c)	(d)	(e)	(f)
Height of column:	6·5 cm		17 cm		15·6	
Depth of water:		210 cm		168 cm		125 cm

(3) The depth on

    (*a*) April 30 was . . .

    (*b*) July 31 was . . .

    (*c*) October 31 was . . .

(4) The depth was

    (*a*) 285 cm on . . .

    (*b*) 96 cm on . . .

    (*c*) 152 cm on . . .

(5) Was July a wet or a dry month? Explain.

(6) During which month was the fall in depth greatest?

(7) The reservoir is full when the depth of water is 400 cm. If its capacity is 520 million litres and the volume of water in the reservoir is proportional to the depth of the water, say how many litres it contains on

    (*a*) March 31, (*b*) September 30, (*c*) December 31.

(8) If restrictions on private use of water come into force when the depth in the reservoir falls below 150 cm, state

    (*a*) the smallest number of litres in the reservoir before the restrictions are imposed,

    (*b*) during which month the restriction was imposed,

    (*c*) during which month the restriction was probably lifted.

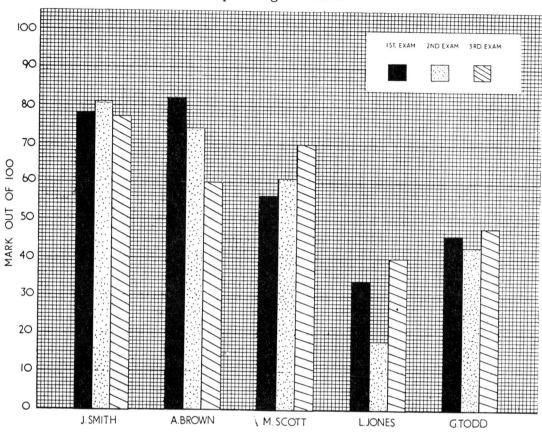
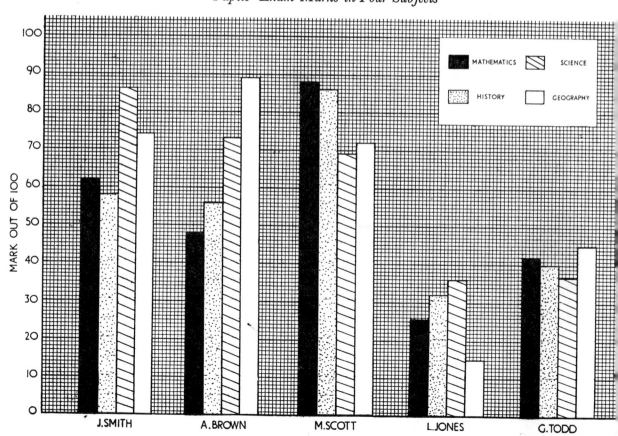

Example 5 ✓

Pupils' English Marks

In this graph you will see how shading the columns differently gives a clear comparison of pupils' performances over 3 exams in the one subject.

Exercises

(1) The scale for marks is 1 cm =

(2) English marks scored were:

    J. Smith   1st exam ....   L. Jones   2nd exam ....

    M. Scott   3rd exam ....   G. Todd   1st exam ....

(3) Give the name of the pupil and the exam in which the following marks were scored:

    (*a*) 40, (*b*) 81, (*c*) 74, (*d*) 56.

(4) Which pupils were most consistent?

(5) Which pupil's marks fell away?

(6) Which pupil showed steady improvement?

Example 6 ✓

Pupils' Exam Marks in Four Subjects

Here shading is also used for comparison, and accentuates the different subject columns.

Exercises

(1) Marks scored were as follows:

    J. Smith   Science ....   A. Brown   Geography ....

    L. Jones   History ....   G. Todd   Mathematics ....

(2) Give the name of the pupil and the subject in which the following marks were scored:

    (*a*) 72, (*b*) 36, (*c*) 56, (*d*) 62.

(3) Say which pupils were

    (*a*) above 50 in all subjects,

    (*b*) below 50 in all subjects.

(4) Which pupil was best in scientific subjects?

(5) Which pupils are above 70 in both science and geography?

(6) Which pupils were most consistent in all subjects?

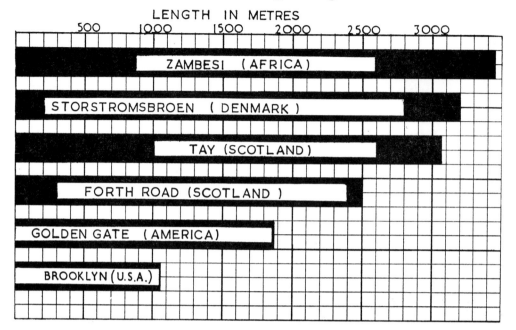

The World's Longest Bridges

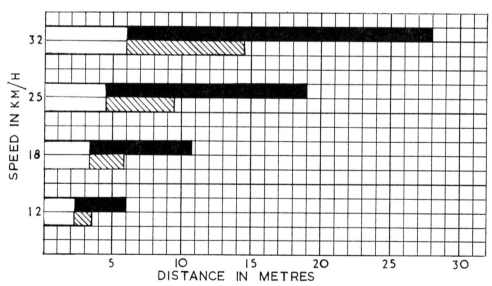

Cycle Stopping Distances on a Level Road

Example 7

The World's Longest Bridges ✓

Sometimes the columns are drawn horizontally; a method which is especially suitable when comparing lengths. The graph shows the lengths of some of the World's Longest Bridges.

Exercises

(1) Which bridges are over 3000 metres in length?

(2) Which two bridges are approximately the same length?

(3) Which bridge is nearly twice as long as Brooklyn Bridge?

(4) Find the lengths of the following bridges,

    (i) Golden Gate, (ii) Forth, (iii) Zambesi.

Example 8

Cycle Stopping Distances on a Level Road

This horizontal column graph helps us to make three comparisons.

(a) How cycle stopping distance increases with speed—by even spacing of columns.

(b) How dry and wet conditions affect stopping distance—by placing columns side by side.

(c) How reaction and braking distances vary—by shading parts of each column.

A shading code is given to help you to answer these questions. WHITE represents 'reaction distance'; SHADED represents 'braking distance (dry)'; BLACK represents 'braking distance (wet)'.

Exercises

(1) Say for what speed on a dry road,

    (a) the braking distance is less than the reaction distance,

    (b) the braking distance is approximately equal to the reaction distance,

    (c) the braking distance is greater than the reaction distance.

(2) Say for what speed on a wet road,

    (a) the braking distance is equal to twice the reaction distance?

    (b) the braking distance is more than 3 times the reaction distance.

(3) Find what is the total stopping distance on a level road which is (a) dry, (b) wet, for the following speeds:

    (i) 12 km/h, (ii) 25 km/h, (iii) 32 km/h.

(4) At what speed is the stopping distance for a wet surface at least twice the stopping distance for a dry surface?

EXERCISES ON DRAWING COLUMN GRAPHS

If you colour your columns they will be more effective. Don't forget to give your graph a title.

(1) The heights in mm of 6 boys in a class were:

Tom	George	Jack	Peter	Stephen	William
1245	1340	1420	1470	1550	1625

Draw a column graph showing these boys in order of size. Use a scale of 1 cm to 100 mm of height, and space your columns evenly.

(2) A column graph was used to show sales of tickets for the school concert by boys and girls of Forms 1, 2, and 3. The final totals were:

Form 1		*Form* 2		*Form* 3	
Boys	Girls	Boys	Girls	Boys	Girls
£26	£22·50	£11·25	£18	£6·50	£10

Use 15 cm to represent £30, and illustrate these figures by a column graph.

(3) A shop's takings for 6 days of a week were as follows:

Mon.	£46	Wed.	£15	Fri.	£25
Tue.	£39	Thurs.	£32	Sat.	£57

Illustrate these by a column graph.

(4) The absences of pupils in a class for 2 weeks, one during an epidemic of influenza, are as follows:

Week ending	Mon.		Tues.		Wed.		Thurs.		Fri.	
	A.M.	P.M.	A.M.	P.M.	A.M.	P.M.	A.M.	P.M.	A.M.	P.M.
5 Nov.	2	2	1	1	2	3	1	2	2	3
12 Nov.	5	6	6	6	8	8	10	11	11	12

By placing graph columns for the two weeks side by side, and by shading or colouring, show how these weeks compare.

(5) Here are exam marks out of 100 for 8 girls in their three arithmetic exams.

	June	Ann	Susan	Grace	Sandra	Wilma	Millie	Joyce
1st exam	58	75	88	34	18	74	84	47
2nd exam	62	48	92	36	22	83	71	47
3rd exam	60	63	95	30	10	86	59	49

Show how these girls are progressing by drawing column graphs in groups of three set side by side (see Example 5). You should use 10 cm to represent 100 marks.

(6) The approximate distances travelled on 1 litre of petrol by cars of various engine sizes are:

Engine Size (c.c.'s)	800	950	1200	1500	1700	2000
Distance (km)	15 to 19	13 to 17	12 to 14	10·5 to 13·5	9 to 12	7 to 10

In drawing your graph place the columns horizontally and shade them from 0 to the shorter distance in kilometres, thus:

(7) The lengths of some of the longest rivers in different parts of the world are given as:

Mississippi	6400 km	Yangtse	5500 km
Amazon	5900 km	Volga	3800 km
Nile	5750 km	Murray	2600 km

Taking 14 cm to 7000 kilometres, you should illustrate these distances by horizontal columns.

Chapter IV

LINE GRAPHS

In drawing column graphs you would see how these columns varied in width from graph to graph. Perhaps you have decided that lines would tell the same story as columns, and could be drawn more quickly.

Because these lines are usually drawn vertically (here meaning up and down the page), they are called *uprights*. In the graphs which follow try to understand

(*a*) how these *uprights* show change,

(*b*) when other *uprights* drawn between the ones in the graphs have a value.

Sometimes, when intermediate uprights have a meaning, by noting the 'trend' in the lengths of the uprights already shown you can guess with reasonable accuracy the lengths of these intermediate uprights. At other times no such estimates can be made. In Example 1 which follows try to estimate the noon temperatures for the 14th, 18th and 22nd of February. When you complete Exercise 2 you will probably find that only your answer for the 18th agrees with the temperature given. Because of the 24-hour time interval between readings wide temperature variations do occur. On the other hand, your answers to Exercises 5 and 6 in Example 2 are likely to be reasonably accurate, since the temperatures at hourly intervals show definite 'trends'.

Note: (1) Temperatures below 0°C are negative, so that 3° below 0°C is written −3°C.

(2) You should copy the graphs in Examples 1, 2 and 4 before answering the exercises.

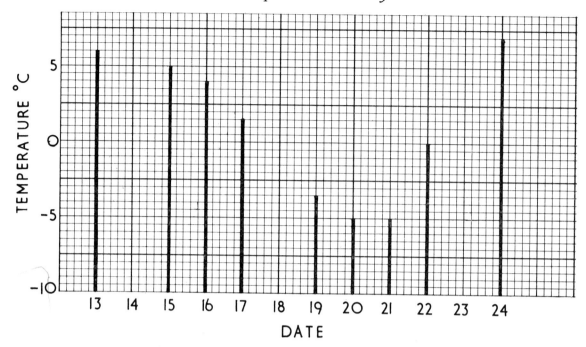

Noon Temperatures in February

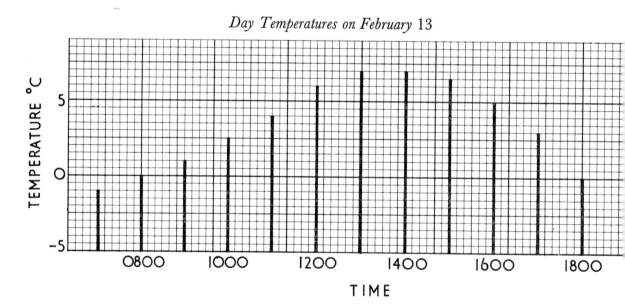

Day Temperatures on February 13

Example 1

Noon Temperatures in February

Exercises

(1) How many degrees Celsius does 1 large division (2 cm) represent on the temperature scale?

(2) Draw **uprights** to show the following noon temperatures:

    (*a*) Feb. 14 7°C,   (*b*) Feb. 18 −1°C,   (*c*) Feb. 23 5°C.

(3) For which dates is the noon temperature below freezing-point (0°C)?

(4) State between which dates the noon temperatures showed

    (*a*) a steady fall, (*b*) a steady rise.

(5) If noon temperatures below 5°C were regarded as belonging to a 'cold spell', between which dates did the cold spell last?

(6) Would intermediate uprights have any meaning? (Note the title of the graph.)

(7) Are the divisions between the dates used here merely to space the uprights evenly?

Example 2

Day Temperatures on February 13

Exercises

(1) State between which hours the temperature

    (*a*) rises steadily, (*b*) falls steadily.

(2) At what times did the highest recorded temperature occur?

(3) Say what the temperature was at:

    (*a*) 1100 hours, (*b*) 1200 hours, (*c*) 1800 hours.

(4) Say at what time(s) the temperature was

    (*a*) 7°C, (*b*) 4·5°C, (*c*) 0°C.

(5) Say, if you drew another upright midway between those for 1100 hours and 1200 hours,

    (*a*) what temperature it would show (approximately),

    (*b*) for what time of day this would be.

(6) Now draw other uprights to show approximate temperatures at

    (*a*) 0730 hours, (*b*) 1030 hours, (*c*) 1520 hours, (*d*) 1645 hours,

and say what these temperatures are.

(7) How many minutes does 1 small division between each hour represent here?

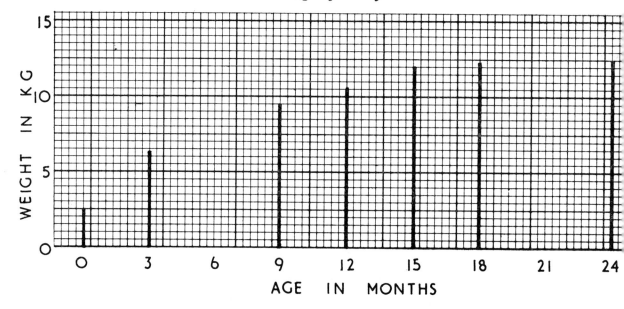

Example 3
Weight of a Baby

Exercises
(1) What scale has been used for weight?
(2) Draw uprights to show the following weights:
   (*a*) at 6 months—8·5 kg, (*b*) at 21 months—12·4 kg.
(3) Give the baby's weight
   (*a*) at birth, (*b*) after 1 year.
(4) Say at what stage the baby's weight increases
   (*a*) most rapidly, (*b*) most slowly.
(5) Say, if you drew an upright midway between 12 and 15 months,
   (*a*) what weight it would show approximately,
   (*b*) at what age this would be.
(6) How many small divisions on the age line show 1 month?
(7) Insert uprights and read off weights at these ages:
   (*a*) 2 months, (*b*) 16 months, (*c*) 22 months.

You should by now have realized that:
(i) There are two quantities in each graph.
(ii) The rise and fall of the uprights show how one quantity changes (varies) as the other is changed (varied).
(iii) Sometimes intermediate uprights can be drawn with reasonable accuracy, sometimes they cannot, and sometimes they have no meaning at all.

The quantities in a graph are often called **variables**—one normally being measured along an upright line, the **vertical axis,** the other a horizontal line, the **horizontal axis.** A separate scale is required for each quantity.

EXERCISES ON DRAWING UPRIGHT LINE GRAPHS

In working these exercises note the value of intermediate uprights.

(1) The number of runs scored by a cricketer in 10 innings is shown as follows:

Innings	1	2	3	4	5	6	7	8	9	10
Runs scored	15	7	4	12	34	42	57	0	33	17

You should space your innings over 10 cm, and use 12 cm to represent 60 runs.
Would intermediate **uprights** have any meaning?

(2) When water was heated and came to the boil after 6 minutes the temperatures in degrees Celsius taken every minute were:

Minutes	0	1	2	3	4	5	6
Temperatures	53	69	82	92	$97\frac{1}{2}$	$99\frac{1}{2}$	100

After drawing your line graph, see if you can insert uprights to tell the approximate temperatures after

$1\frac{1}{2}$, $2\frac{3}{4}$, $4\frac{1}{4}$, and $5\frac{1}{2}$ minutes.

(3) The height of an aeroplane in metres, noted every 30 minutes, was:

Time	0800	0900	0930	1000	1030	1100	1130	1200
Height	1200	1350	1450	1400	860	900	700	300

Draw a line graph spacing your uprights 1 cm apart, and using a scale of 1 cm to represent 100 m in height.

Say if intermediate uprights could be drawn to find the approximate height of the plane at:

(a) 0845 hours, (b) 1015 hours, (c) 1106 hours.

Would additional information help you to draw these uprights?

Chapter V
JAGGED LINE GRAPHS

From the rise and fall in the uprights which formed most **line** graphs we were able to see how changes in one quantity followed changes in another. Now by joining the tops of the uprights and then omitting the uprights themselves we form **jagged line** graphs which tell us at a glance something about the speed of these changes by the steepness of rise and fall.

The quantity for which we choose values is measured along the **horizontal axis.** The other quantity, which depends on the values chosen, is measured up the **vertical axis.** In a graph on day temperatures we choose suitable times at which to read the thermometer, and the temperatures will depend on the times chosen. Thus 'Time' is measured along the horizontal axis and 'Temperature' is measured up the vertical axis.

From the following graph topics, write down the quantity which should be measured along the horizontal axis:

(1) Average weights of girls of various ages.
(2) Day of month and lighting-up times for vehicles.
(3) Cost of article and total value of sales.
(4) Demand for gas at various times of day.
(5) Weights of letters and postage.
(6) Stopping distance of car for various speeds.
(7) Distance from Dover and varying depth of English Channel.
(8) Length of spring stretched by various weights.
(9) Boy's age in years and his weight in kg.
(10) Life Assurance premium and age next birthday.

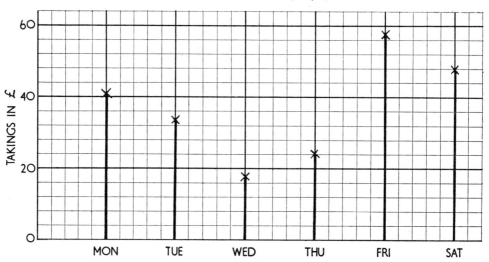

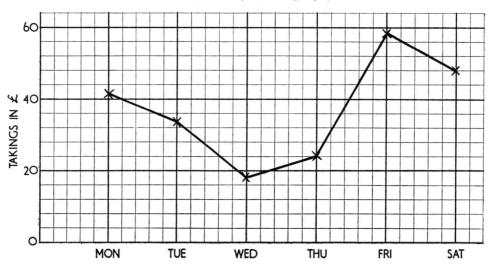

Examples 1A *and* 1B

A Shop's Daily Takings

Notice first how the crosses which marked the tops of the uprights in the upper graph have been joined in the lower graph to form a jagged line. Now cover the upper graph and use only the lower graph to answer the following questions:

Exercises

(1) Give the takings on:

   (*a*) Wednesday, (*b*) Saturday.

(2) Say on which days the takings were:

   (*a*) £34, (*b*) £58.

(3) From the previous day, find which day's takings showed

   (*a*) the steepest rise, (*b*) the least drop.

(4) Which day was probably the half-holiday?

(5) Account for:

   (*a*) The high takings on Friday.
   (*b*) The small increase in takings from Wednesday to Thursday.

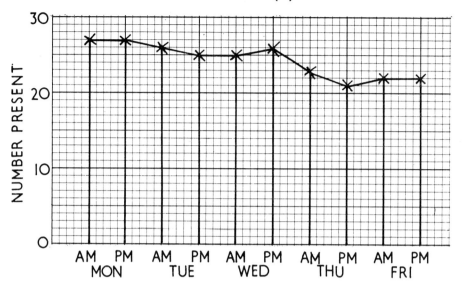

Class Attendances (A)

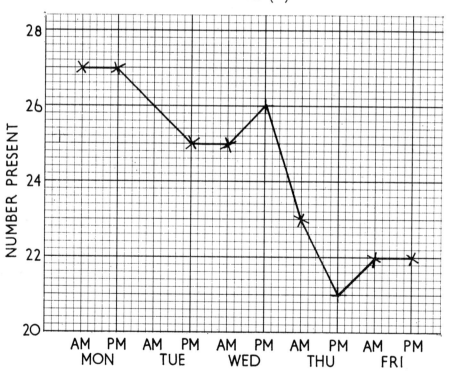

Class Attendances (B)

Examples 2A and 2B

Class Attendances

In comparing attendance and absence graphs in Chapter III for the same class we saw how the larger numbers forced us to use a smaller scale in the attendance graph. In Graph A opposite you will notice the dotted line through the uprights, drawn to show a part of each upright which is common to all attendances. With this common part omitted in Graph B, the attendance scale has been increased, and changes in attendance are shown by steeper rising and falling of the graph line.

Exercises

(1) Give the lowest number on the vertical axis

   (*a*) in Graph A, (*b*) in Graph B.

(2) Each pupil present is shown on the vertical axis by a line

   (*a*) . . . cm long in Graph A.

   (*b*) . . . cm long in Graph B.

(3) So the scale in Graph B is . . . times as large as the scale in Graph A.

   Now use Graph B only.

(4) (*a*) How many days show the same P.M. as A.M. attendance?

   (*b*) What type of line shows this?

(5) Which day shows a rise in attendance in the afternoon?

(6) Say which day shows the sharpest drop in attendance:

   (*a*) in the afternoon from the morning,

   (*b*) in the morning from the previous afternoon.

(7) Will intermediate points on the graph-line have any meaning?

N.B. In Graph B, because the vertical scale does not start at zero, the attendance shown by any point on the graph is **not** in proportion to the height of the graph above the horizontal axis at that point. *E.g.*, the attendance on Wednesday P.M. is **not** 3 times the attendance on Friday P.M. Comparisons may only be made from values on the vertical axis.

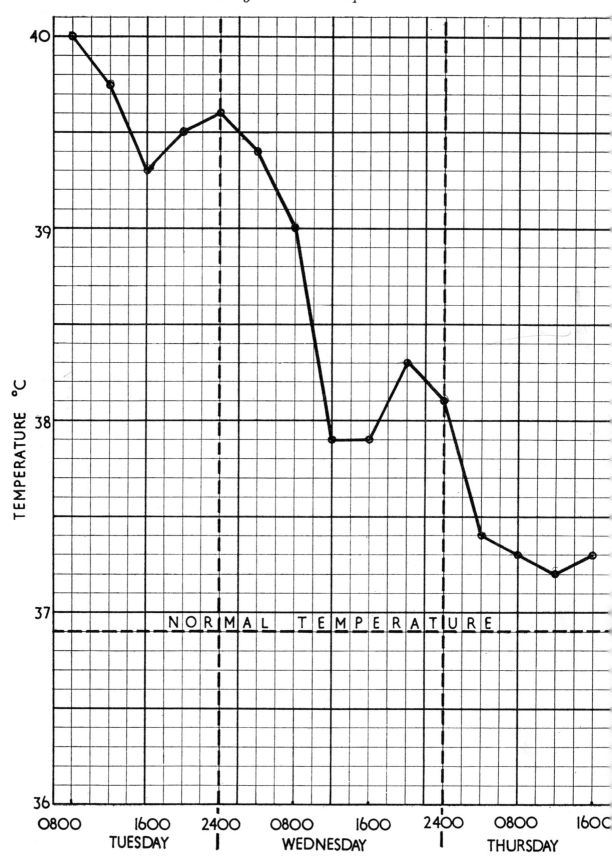

Example 3

John Smith's Temperature

In this graph a large part of the temperature scale has been omitted because a boy's temperature is unlikely to fall below 36° Celsius.

Exercises

(1) What is the lowest temperature shown on the **vertical axis**?
(2) The scale for temperature is ... cm = 1°C, ... cm = 0·1°C.
(3) The scale for time on the **horizontal axis** is ... divisions = 1 hour.
(4) According to the graph, how often was John Smith's temperature taken?
(5) Give the times between which his temperature was falling

   (*a*) on Tuesday, (*b*) on Wednesday.

(6) Give the times between which his temperature was rising

   (*a*) on Tuesday, (*b*) on Wednesday.

(7) At what successive times was his recorded temperature the same?
(8) Give John Smith's temperature at

   (*a*) 0800 hours on Tuesday, (*b*) 2000 hours on Wednesday, (*c*) 0400 hours on Thursday.

(9) What is **normal temperature**?
(10) How much above normal was John's temperature at 0400 hours on Thursday?
(11) Give his temperature approximately

   (*a*) 0600 hours on Wednesday, (*b*) 1800 hours on Wednesday.

(12) John was given drugs from 0800 hours on Tuesday. When did these drugs cause his temperature to drop sharply?

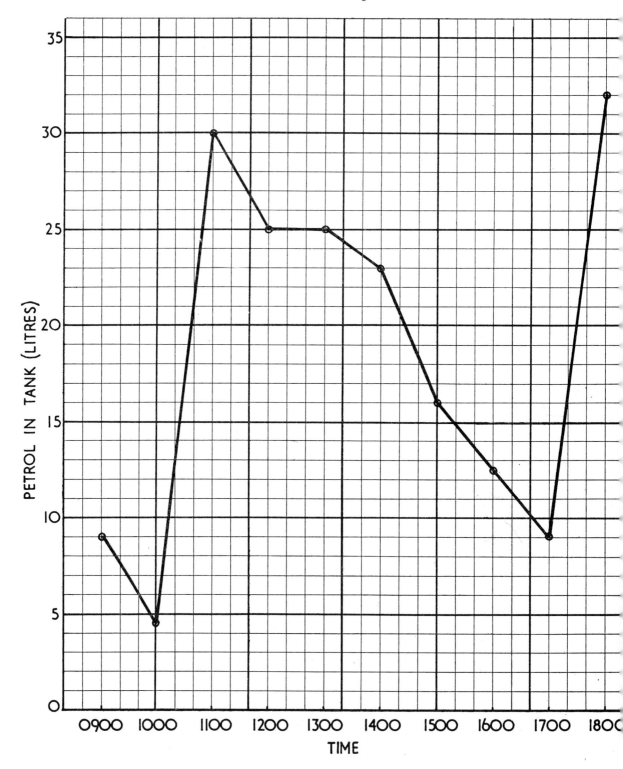

Petrol in Tank of a Car

Example 4

Petrol in Tank of a Car

The driver of a car notes from the petrol gauge every hour how many litres are in the car-tank. The tank holds 34 litres, and the car averages 12 km to the litre. You will find here how intermediate points on the graph line take the place of intermediate uprights.

Exercises

(1) Say what height on the **vertical axis** represents

    (*a*) 5 litres, (*b*) 1 litre.

(2) Say what length on the **horizontal axis** represents

    (*a*) 1 hour, (*b*) 20 minutes, (*c*) 3 hours 40 minutes.

(3) Say how many litres of petrol were in the tank

    (*a*) at the start, (*b*) at 1100 hours, (*c*) at 1600 hours, (*d*) at 1700 hours, (*e*) at 1630 hours.

(4) Say at what time of day (when the car was running) there was in the tank

    (*a*) 4·5 litres, (*b*) 24 litres, (*c*) 15·5 litres.

(5) Between what times was the tank filled up?

(6) Say how far, approximately, the motorist travelled between

    (*a*) 0900 and 1000 hours, (*b*) noon and 1400 hours, (*c*) 1500 and 1700 hours.

(7) If the motorist travelled only 30 kms between 1000 and 1100 hours, how many litres of petrol were put in the car?

(8) At high speed the car travels only 9 km per litre. Approximately how far did it go between 1400 and 1500 hours, when its speed was high?

(9) Find approximately how many litres were in the tank at

    (*a*) 1130 hours,

    (*b*) 1430 hours,

    (*c*) 0945 hours.

(10) Between what times can you not read from the graph the approximate amount of petrol in the tank?

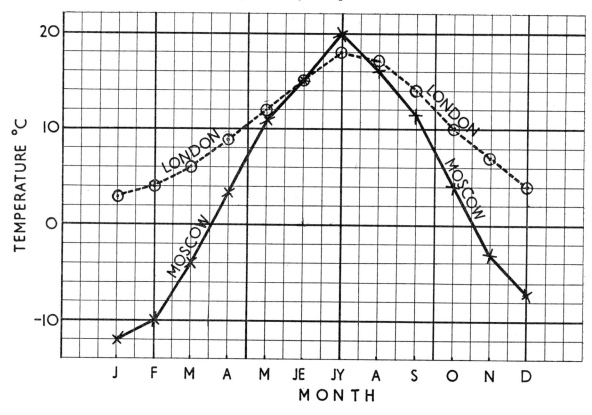

Average Monthly Temperatures

Example 5

Average Monthly Temperatures
(London and Moscow)

Jagged line graphs drawn on the same sheet are useful for making comparisons.

Exercises

(1) Which town is usually the colder?

(2) Say during which season of the year the differences of temperature in London and Moscow are:

(*a*) greatest, (*b*) least.

(3) Which town shows the sharper changes of temperature?

(4) Give the temperatures of these towns in

(*a*) February, (*b*) July, (*c*) September.

(5) Find when the average London temperature is

(*a*) 17°C, (*b*) 4°C.

(6) Find when the average Moscow temperature is

(*a*) −12°C, (*b*) 11°C.

(7) Do the intermediate points on these graphs have any meaning?

(8) Say for which months the Moscow temperatures are below freezing-point.

(9) What is the greatest number of degrees below freezing-point shown for any month in Moscow?

(10) How many degrees above freezing-point is the average temperature in London for the same month?

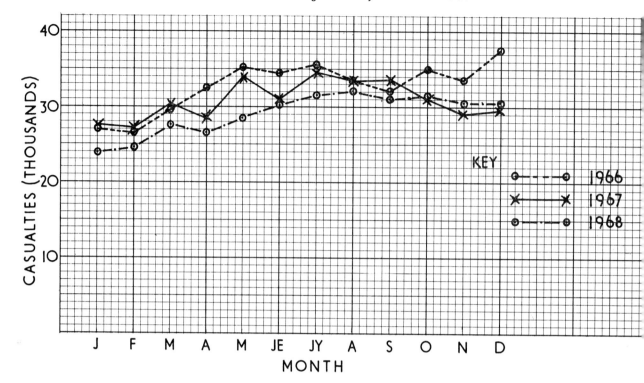

Example 6

Road Casualties Month by Month
(1966, 1967 and 1968)

In this graph you should reason out which upward or downward movements are caused by seasonal factors, and as you answer the exercises, which are affected by other events. Before reading the graph study the code (or key) to the lines which represent the casualties for the different years.

Exercises

(1) For which months of all three years were the casualties below 28,000? Can you suggest why these months give the lowest figures?

(2) Say for how many months the 1967 casualties were

   (*a*) more than the 1966 casualties,

   (*b*) less than the 1966 casualties,

   (*c*) approximately equal to the 1966 casualties.

(3) The 'Breathalyser Test' was first introduced in October 1967. Explain how its effect on casualties is shown on the graphs.

(4) State any other factor(s) which might have caused the 1967 casualty rates to be lower in general than those of 1966, despite the increase in the number of vehicles registered in 1967.

(5) Why do the graph lines rise in summer?

(6) Give the approximate numbers of casualties in each of the three years for

   (*a*) March, (*b*) June, (*c*) November.

(7) Copy this graph on a scale for casualties of 2 cm to 5000. Note how your graph illustrates more clearly a rise or fall in the number of casualties from month to month.

EXERCISES ON DRAWING JAGGED LINE GRAPHS

In working these exercises you should not draw in any uprights but only mark their top points. Do this with either a thin cross or a fine dot surrounded by a circle.

(1) Here are John's total savings at the end of each month of last year:

Month	Jan.	Feb.	Mar.	Apr.	May	June	July	Aug.	Sept.	Oct.	Nov.	Dec.
Savings (£)	0·70	1·10	1·30	1·05	1·45	1·80	1·90	0·60	0·75	0·95	1·25	1·35

You should use 10 cm to £2·00.

(2) Draw a jagged line graph to show the falling off in the average number of pupils taking school lunches each day.

Month	Sept.	Oct.	Nov.	Dec.	Jan.	Feb.	Mar.	Apr.	May	June
Average number taking lunches	252	216	204	196	190	186	194	148	142	140

Start your vertical axis number at 140.

Will intermediate points on your graph line have any meaning?

(3) The weight of a pig is given from birth until the end of the 16th week.

Week	0	1	2	3	4	5	6	7	8	9	10	11	12	13	14	15	16
Weight (in kg)	1	2	3·5	5·5	7·5	10·5	13·5	17	20	18	21	25	28·5	32	34	38	42·5

Use 16 cm for the **horizontal axis** and 10 cm for the **vertical axis.** From points on this graph try to guess the pig's weight after:

(a) $6\frac{1}{2}$ weeks, (b) $14\frac{1}{2}$ weeks.

(4) Compare the percentage of absences for boys and girls in a school over the 11 weeks given here.

Month	October				November				December	
Week ending	5	12	19	26	2	9	16	23	30	7	14
Boys' percentage	3·4	4·0	4·8	6·6	8·4	8·5	12·7	6·0	4·7	9·1	5·4
Girls percentage	3·0	3·2	4·0	9·1	11·3	15·4	9·6	7·3	5·1	8·5	6·7

Here you will need 16 cm for percentage absence on the vertical axis.
Will intermediate points on these graphs have any meaning?

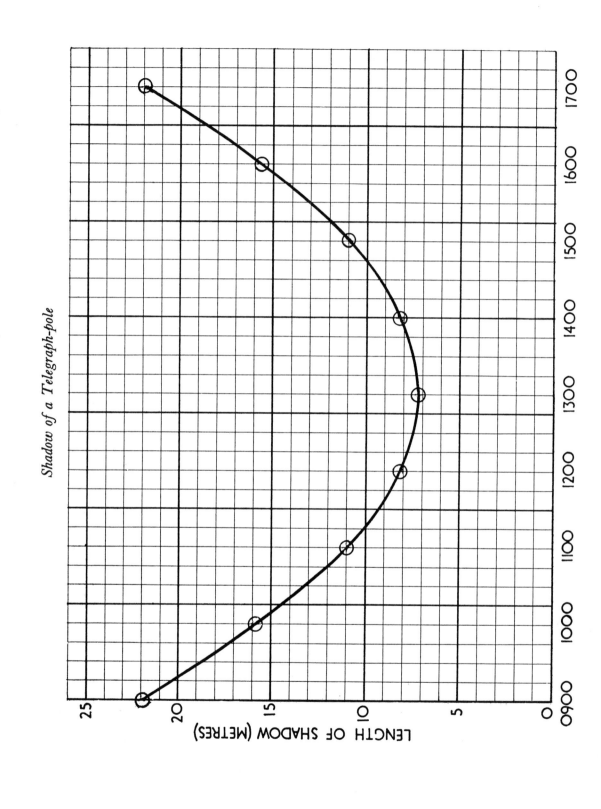

Chapter VI
CURVED GRAPHS

When a patient's temperature is recorded every four hours the graph is usually shown as a jagged line. This is all that doctors require to see how the patient is progressing. But in fact, as time passes, the temperature changes gradually, and should be shown by a smooth curve through the points marked. You will understand this best if you can watch a barograph—an instrument that records air pressure—on graph paper which is rotating steadily by clockwork on a cylindrical drum. Perhaps there is one in your school science laboratory.

In curved graphs intermediate points will have a meaning, and their values can be read more or less accurately.

Example 1

Shadow of a Telegraph-pole

As time passes the shadow of the telegraph-pole alters gradually, and so when the points marking the length of this shadow are joined they form the smooth curve shown. Notice how this one is evenly balanced (symmetrical) about a line through its lowest point.

Exercises

(1) For which of the quantities, time of day or length of shadow do we choose suitable values? This is the independent quantity.

(2) On the "time of day" scale, what does 1 small division (0·5 cm) represent?

(3) Say at what time(s) the shadow is

    (*a*) longest, (*b*) shortest.

Explain.

(4) Say between what two hours the shadow changes

    (*a*) most rapidly, (*b*) most slowly.

(5) Find the length of the shadow at

    (*a*) 1000 hours, (*b*) 1600 hours, (*c*) 1700 hours, (*d*) 1430 hours.

(6) State the time(s) at which the shadow length is

    (*a*) 11 m, (*b*) 7·2 m, (*c*) 17·5 m.

(7) Explain why the shadow is shortest at 1300 hours, and not at noon.

Distance of the Horizon

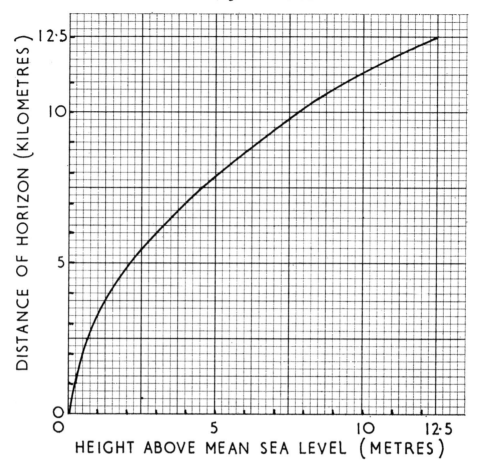

A Barograph

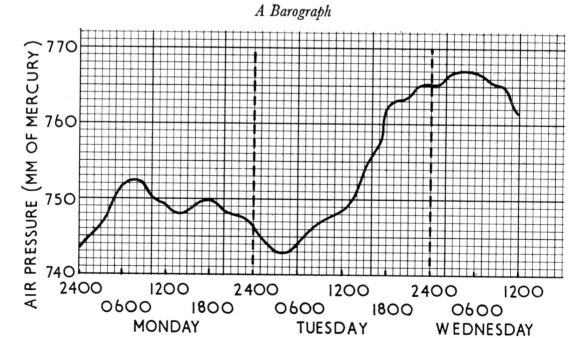

Example 2

Distance of the Horizon

This graph shows how far you can see to the horizon over a calm sea for various heights above sea-level.

Exercises

(1) The scales are:
   (a) for height above sea-level . . . cm = 5 m, therefore . . . mm = 1 m,
   (b) for distance of horizon . . . cm = 5 km, therefore . . . mm = 1 km.
(2) Say for which increases in height the horizon distance
   (a) increases most rapidly, (b) increases most slowly.
(3) Give the horizon distance
   (a) for a height of 2 m, (b) for a height of 8·25 m.
(4) Find the height of the observer if the horizon distance is
   (a) 7·0 km, (b) 12·0 km.
(5) What is the increase in the horizon distance for an observer who climbs from a height of 4·25 m to a height of 10·5 m?
(6) To increase the horizon distance from 8·0 km to 11·0 km an observer must climb from . . . m to . . . m above sea-level.

Example 3

The Barograph

This is still a curved graph, and although the pen marking the curve constantly changes its direction, it traces a smooth curve.

Exercises

(1) (a) On the horizontal axis 0·2 cm = . . . hour.
   (b) On the vertical axis 0·2 cm = . . . mm of mercury.
(2) State the day and time when
   (a) the highest pressure was recorded, (b) the lowest pressure was recorded.
(3) Say between what times on Tuesday the pressure
   (a) rose most steeply, (b) fell most rapidly.
(4) Give the air-pressure
   (a) at 1200 hours on Monday, (b) at 1700 hours on Tuesday, (c) at 0100 hours on Wednesday.
(5) Find the time(s) when the air-pressure was
   (a) 744 mm, (b) 762 mm, (c) 752·4 mm.

Note: You should find out from your science teacher how the changes in air-pressure shown on the drum of a barograph help you to forecast the weather.

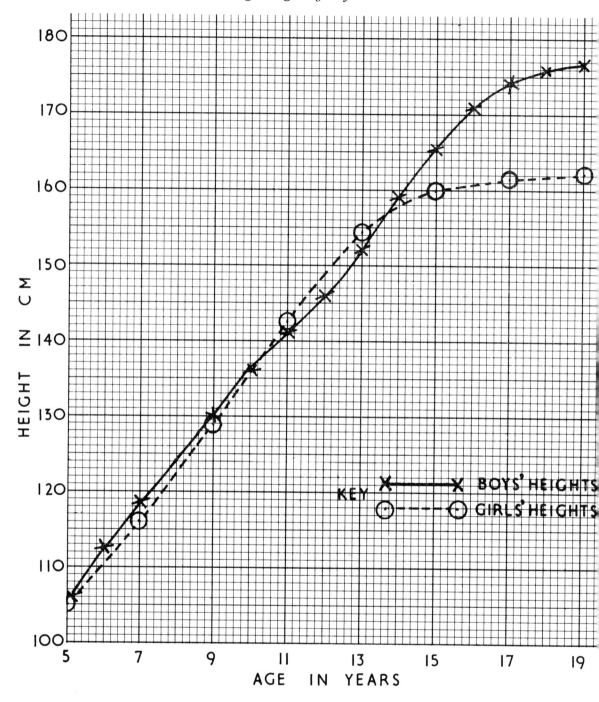

Average Heights of Boys and Girls

Example 4

Average Heights of Boys and Girls

Heights of boys and girls also change smoothly, and these comparison graphs will show interesting differences in the way they grow.

Exercises

(1) The range of values shown on the axis for

    (*a*) **Age** is from ... to ... = ... years.

    (*b*) **Height** is from ... to ... = ... cm.

(2) (*a*) ∴ 7 large divisions = ... years gives a scale of 1 large division = ... years.

    (*b*) ∴ 8 large divisions = ... cm of height gives a scale of 1 large division = ... cm of height.

(3) At what ages is the girls' average height greater than the boys'?

(4) Say after what age the height increase slows down

    (*a*) for girls, (*b*) for boys.

(5) (*a*) At what ages are their heights the same?

    (*b*) What are their heights then?

(6) Find the boys' average height at

    (*a*) 8 years, (*b*) 12 years, (*c*) 15 years, (*d*) $17\frac{1}{2}$ years.

(7) Find the girls' average height at

    (*a*) 8 years, (*b*) 12 years, (*c*) 15 years, (*d*) $17\frac{1}{2}$ years.

(8) Find at what age the boys' average height is

    (*a*) 130 cm, (*b*) 162·5 cm.

(9) Find at what age the girls' average height is

    (*a*) 145·5 cm, (*b*) 161 cm.

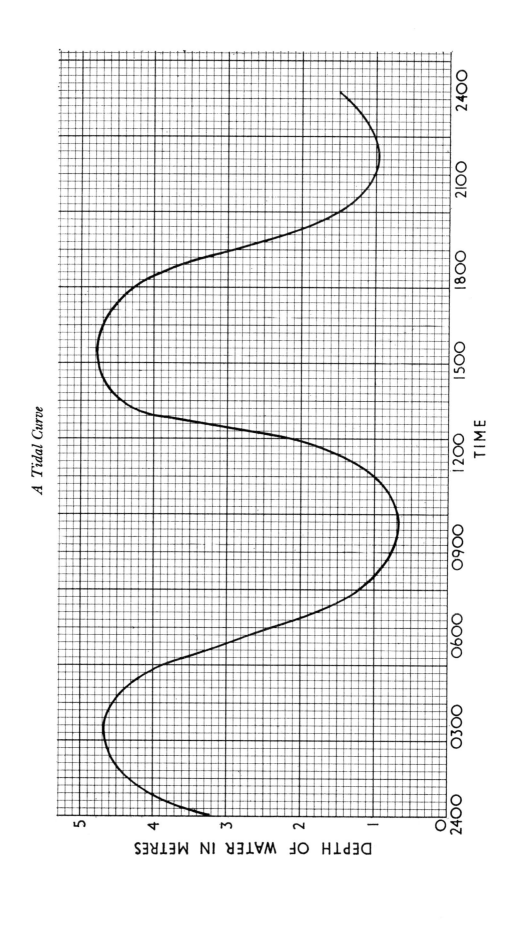

A Tidal Curve

Example 5

A Tidal Curve

The graph opposite shows the depth of water in a tidal river over 24 hours.

Exercises

(1) (a) The scale for time is 2 mm = . . . minutes.
    (b) The scale for depth of water is 2 mm = 0· . . . metres.

(2) (a) The first high tide was at approximately . . . hours.
    (b) The depth of water then was . . . m.

(3) (a) The second high tide was at approximately . . . hours.
    (b) This was . . . hours . . . minutes after the first.

(4) State the depth of water at the following times:

    (a) 0100 hours, (b) 0600 hours, (c) 1200 hours, (d) 1830 hours, (e) 1945 hours.

(5) Say at which times of day, approximately, the depth was

    (a) 2·5 metres, (b) 4·3 metres.

(6) If James King required 3·6 m of water to launch his boat, state the periods of time between which this would be possible.

Amounts at Simple and Compound Interest

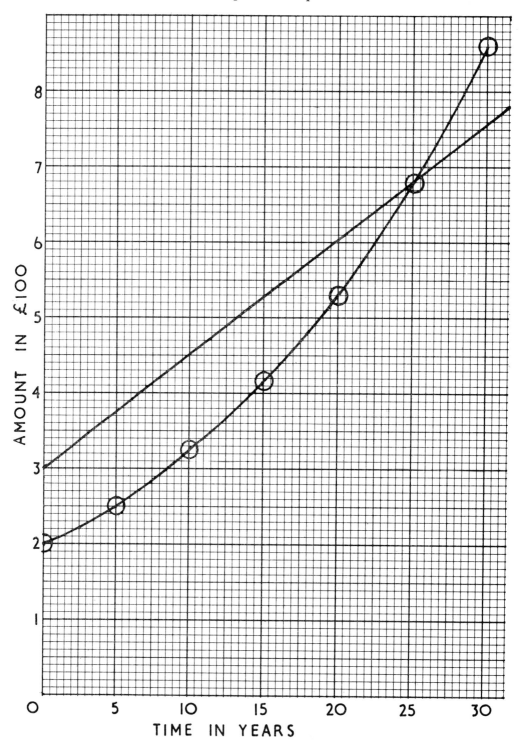

Example 6

Amounts at Simple and Compound Interest

The graphs on the opposite page show the growth over periods of up to 30 years of

(a) £200 invested at 5% per annum compound interest, and

(b) £300 invested at 5% per annum simple interest.

Exercises

Use the graphs to answer the following:
(1) Find the amount at 5% compound interest of £200 after
    (a) 12 years, (b) 26 years.
(2) Find the compound interest on £200 at 5% after
    (a) 8 years, (b) 16 years.
(3) Say after (approximately) how many years at 5% compound interest the principal is
    (a) doubled, (b) trebled.
(4) After how many years at 5% simple interest is the principal doubled?
(5) State the approximate difference between the two investments after
    (a) 16 years, (b) 22 years.
(6) State after how many years the two investments will be approximately equal.
(7) Find the approximate amount at 5% compound interest of
    (a) £100 after 18 years, (b) £1000 after 7 years.
(8) Find the amount at $7\frac{1}{2}$% simple interest of
    (a) £600 after 20 years, (b) £300 after 14 years.

Scales

By this time you should have realized that the quantity measured along each axis has to be spaced out evenly, so a scale is necessary. The scale will depend on (i) the range of values, and (ii) the number of large divisions available. It should normally be as large as possible, *i.e.*, there should be as few units to each large division as possible.

You will also have noticed that there are two types of squared paper used for graph work. Paper A has large divisions each 2 cm subdivided into 10 small divisions each 0·2 cm (or 2 mm). Paper B has large divisions each 2·5 cm subdivided into 5 small divisions each 0·5 cm. To make the best use of the 10 small divisions to each large division using paper A, the number of units to each large division should either be

(1) a *factor* of 10, such as 1, 2, $2\frac{1}{2}$, 5, 10, so that each unit is shown by an exact number of small divisions. For example, if 5 units = 10 small divisions, then 1 unit = 2 small divisions,

or

(2) a *multiple* of 10, such as 20, 30, 40, so that each small division represents an exact number of units. For example, if 10 small divisions = 30 units, then 1 small division = 3 units.

For paper B the number of units to each large division should either be
(1) a *factor* of 5 such as 1, $2\frac{1}{2}$, 5,
or (2) a *multiple* of 5 such as 10, 15, 20, 25.

Paper A measures 12 large divisions by 9 large divisions and is preferred where greater accuracy is required in drawing or in reading a graph.

Paper B measures 9 large divisions by 7 large divisions.

To arrive at the scale figure using paper A:

(i) Divide the range of values by the number of large divisions.

(ii) Choose the factor or multiple of 10 equal to or greater than your quotient.

E.g., (a) If the range is one of 52 weeks

12 large divisions = 52 weeks ∴ 1 large division = $4\frac{1}{3}$ weeks.

Choose 1 large division = 5 weeks and then 1 week = 2 small divisions.

or (b) If the road level varied from 5 metres (above mean sea-level) to 252 metres the range is 247 metres, and 12 large divisions = 247 cm ∴ 1 large division = 20+ m. Choose 1 large division = 30 m and then 1 small division = 3 m.

Note: Using the 9 large divisions instead of 12 for the range of 247 m will give the same scale in this case.

When drawing curved graphs draw the lines as lightly as possible to begin with, as you may have to smooth out kinks, and draw from the inside of the curve. In some cases curved graphs are easier to draw if the scale is not too large, but remember that if you make an error of 1 small division with a scale of 1 large division = 20 units, your error will be 2 units, while with a scale of 1 large division = 10 units your error will only be 1 unit.

Exercise

(1) From the following table choose the most suitable scales for use with paper A. You need not use all the divisions available if a smaller number would give an equally good scale.

	Number of large divisions available	*Range of values*
(a)	7	0–35 pupils
(b)	9	0–18 metres
(c)	5	£0–£85
(d)	12	40–160 km
(e)	9	64–73 degrees
(f)	7	0–30 kg
(g)	8	42–76 eggs
(h)	7	32–212 accidents
(i)	12	25–65 days
(j)	9	£80–£365
(k)	8	4·5–18·3 cm
(l)	7	483–1010 litres

(2) Repeat all examples except (d) and (i), assuming that paper B is used.

EXERCISES ON DRAWING CURVED GRAPHS

(1) If you could use a sling to send a stone vertically upward at 44 metres per second its height at various times would be:

Time (s)	0	$\frac{1}{2}$	1	$1\frac{1}{2}$	2	3	4	5	6	7	$7\frac{1}{2}$	8	$8\frac{1}{2}$	9
Height (m)	0	21	39	55	68	88	98	98	88	68	54	38	20	0

Draw a curved graph through the points, and find

(a) its height after $1\frac{1}{4}$ seconds,

(b) when its height was 78 m,

(c) what was the greatest height reached and the time then.

(2) The area of squares with different lengths of sides is shown thus:

Length of side (cm)	0	1	$1\frac{1}{2}$	2	3	$3\frac{1}{2}$	4	5	$5\frac{1}{2}$	6
Area of square (cm²)	0	1	$2\frac{1}{4}$	4	9	$12\frac{1}{4}$				

Complete the table, choose your scales, and draw the graph. See if you can find from the graph

(a) the area of a square of side 4·3 cm,

(b) the length of side of a square of area 7·3 cm².

(3) The average weights of selected groups of boys and girls of different ages are:

Age	3	5	7	9	11	13	15	17
Boys' weight (kg)	16	19	23·2	27·7	34·3	41·8	50·5	62·7
Girls' weight (kg)	15·5	18·2	22·2	27·1	34·1	43·5	50·5	58·6

Draw the graphs, and find out

(a) when the girls' average is greater than the boys',

(b) at what ages (approximately) their weights are the same,

(c) if their weights appear to be reaching their limit at 17.

(Your weight axis may be marked from 15 to 65 kg.)

(4) The following table gives the *amount* to which £100 accumulates at 4% per annum compound interest after a given number of years.

Number of years (n)	0	5	10	15	20	25	30	35
Amount of £100 (A)	100	122	148	180	219	266	324	395

(a) Use this table to find the amount at 4% per annum compound interest on

(i) £400 after 20 years, (ii) £150 after 30 years.

(b) Using a scale of 2 cm to 5 years along the horizontal axis and 2 cm to £50 along the vertical axis, draw a graph to show how A varies with n.

(c) Use your graph to estimate to the nearest £5 the compound interest on

(i) £300 after 18 years, (ii) £250 after 24 years.

(d) On the same graph sheet, draw a graph to show the amount of £100 at 4% per annum *simple* interest for periods up to 30 years.

(e) Use your graphs to estimate the smallest number of years which must elapse before the amount of £100 at 4% compound interest exceeds the amount of £100 at 4% simple interest by
(i) £40, (ii) £75.

(5) The following table gives the amount of capital repaid on a mortgage of £1000 repayable over 25 years.

Year	1	5	10	15	20	25
Amount of capital repaid	21	114	261	435	690	1000

(a) State how much capital is still to be paid after
(i) 15 years, (ii) 20 years.

(b) State how much capital of a mortgage of (i) £1500 has been repaid after 10 years; (ii) £2500 has been repaid after 15 years.

(c) Using a scale of 2 cm to 5 years horizontally and a scale of 2 cm to £100 vertically, draw a graph to show the variation of the amount of capital repaid with the number of years.

(d) Use your graph to find the amount of capital on a mortgage of £2000 still to be repaid after
(i) 12 years, (ii) 16 years, (iii) 21 years.

(6) The following table shows the effect of using different proportions of soil and 'Shoot', a new fertilizer, on the growth of bean plants after 7 weeks:

No. of g of 'Shoot' per kg of soil	0	4	8	12	16	20	24	28	32
Height of bean plants (cm)	6·4	9	10·2	10·7	10·6	9·8	8·5	6·9	5·6

(a) Show this on a graph, using 1 cm to represent 2 g on one scale and 1 cm to represent 1 cm of height on the other.

(b) How much 'Shoot' should be mixed with each kg of soil to give the greatest growth?

(c) At above what number of g per kg of soil does 'Shoot' appear to hinder growth?

(d) How many g of 'Shoot' per kg of soil would you use to increase the height of the bean plants by
(i) 40% above normal, (ii) 60% above normal.

(7) A farmer has 55 m of fencing with which to form two sides AB, BC of a rectangular sheep-pen in the corner of a field (see sketch). Complete the following table, and draw a graph showing how the area varies with the length of AB. Use the graph to find (i) the largest area he can enclose; (ii) the lengths of AB and BC which produce this area.

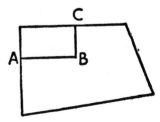

Length of AB (in m)	0	10	20	25	30	35	40	50	55
Length of BC (in m)	55	45	35						
Area of pen (in m²)	0	450	700						

(8) The heights of a clay pigeon taken every half second are shown in the table:

Time in seconds	0	0·5	1	1·5	2	2·5	3	3·5	4
Height in metres	0	8·5	14·5	18·5	19·5	18·5	14·5	8·5	0

Using scales of 2 cm to represent 0·5 seconds on the horizontal axis and 5 m on the vertical axis, plot the graph of these readings.

Using your graph, answer the following questions:

(a) The marksman hits the target 1·7 seconds after release. At what height was the target?

(b) The marksman prefers to hit the target when it is 17 m above the ground. At what times after release can he hit the target?

(c) The marksman can only fire at the target when it is 10 m or more above the ground. For how long a period can the marksman fire at the target?

(9) The table shows the population of a town over a period of years.

Year	1940	1945	1950	1955	1960	1965	1970
Population	92 200	92 400	92 700	93 100	93 650	94 300	95 150

Show these figures on a graph and use your graph to estimate

(1) the population in

(a) 1954, (b) 1958, (c) 1967,

(2) in which year the population reached

(a) 94 000, (b) 95 000

(10)
Time in years	1	2	3	4	5
Amount	£110	£121	£133·10	£146·40	£161·05

The above table shows the amount after a given time for £100 invested at compound interest.

Using scales of 2·4 cm to represent 1 year and 2 cm to represent £10, draw a smooth curve to illustrate this table.

(a) State the rate of interest.

(b) From your graph estimate:

(i) the time when the amount is £130 giving your answer to the nearest month,

(ii) the amount after $3\frac{1}{2}$ years giving your answer to the nearest £,

(iii) the amount after 6 years giving your answer to the nearest £.

Chapter VII

STRAIGHT LINE GRAPHS

When two related quantities are such that when one is doubled this doubles the other, when one is trebled this trebles the other, and so on, these quantities are in direct proportion.

E.g., a number of articles and their total cost,
or a number of bricks and their total weight.
Can we discover what kind of graph to expect between these two quantities?

If the 'rise' of each step on a stair is 15 cm, while the 'tread' is 25 cm, the height risen and the horizontal distance travelled are in direct proportion. A sketch of 3 steps should show us what kind of graph to expect.

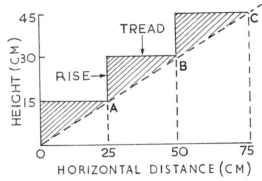

For a distance of 25 cm, height risen is 15 cm, so the first point on the graph is A.
For a distance of 50 cm, height risen is 30 cm, so the second point on the graph is B.
Similarly, the third point on the graph is C.
Do the lines from A to B and from B to C continue at the same slope as the line from the **origin** (where the axes meet) to A? The equal triangles above the lines OA, AB, and BC representing steps will show that they do, like the slope of a banister. So the graph connecting two quantities in proportion will be, like OABC, a **straight line.**

The straight line graph is most useful as a ready reckoner where a large number of results, instead of being calculated, can be read from a graph.

Additional sections connected with each graph which are to be found at the end of this chapter are intended only for pupils studying algebra.

Discount at 6½p in the £

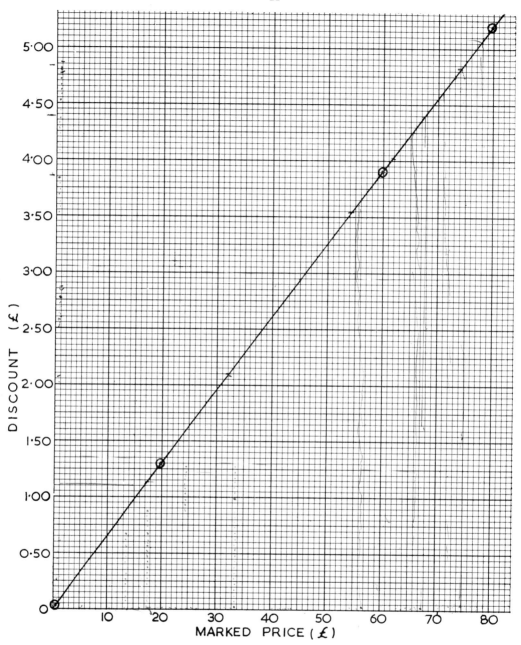

Example 1

Discount Graph

Find the discount at 6½p in the £ on Marked Prices (M.P.) up to £80.
The discount is in proportion to the M.P. in £'s, so that the graph will be a straight line. In this type of graph you must first make up your own table of values. To be sure of obtaining the correct straight line at the first attempt 4 values should be calculated.[1]

M.P. (£'s)	0	20	60	80
Discount (£'s)	0	1·30	3·90	5·20

Plotting these values gives the graph opposite. Use the graph to read off the following values.

Exercises

(1) The discount for the following marked prices:
£6, £14, £18, £27, £34, £45·50, £57, £67, £73, £78·50.

(2) For what marked prices the discount is:
65p, £1·10½, £2·08, £2·60, £2·86, £3·51, £4·03, £4·55, £4·81, £4·94.

(3) The discount for these marked prices (by halving and doubling):
£120, £136, £150.

(4) For what marked prices the discount at 13p in the £ would be:
£1·56, £3·70, £6·50.

(5) Using the same scales as in this graph, draw a graph showing the dividend at 4½p in the £ on purchases at a Co-operative Store up to £100. Now set one of your classmates questions like the ones in the Discount Graph.

[1] If the position of two points on a straight line graph are known they can be joined to form the correct graph. But if one point is wrongly plotted the resulting graph will be wrong.

The Length of a Spring

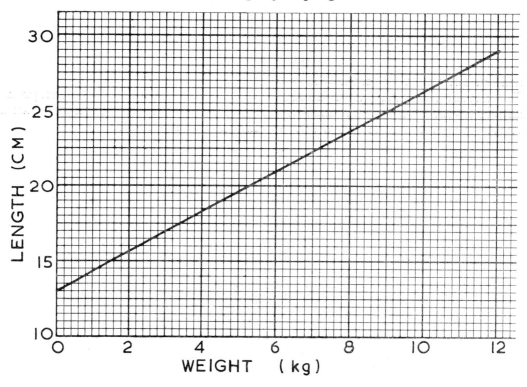

A Knitting Conversion

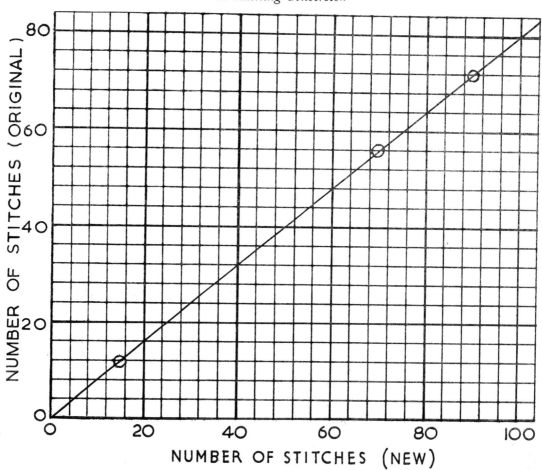

Example 2

The Length of a Spring

A spiral spring suspended from one of its ends has various weights attached in turn to the other end. The increase in length is in direct proportion to the weight attached, within the elastic limit of the spring. The results of an experiment with a spiral spring are:

Weight (kg)	1	4	7	9	12
Stretched length (cm)	14·7	18·6	22·5	25·1	29·0

The points are plotted and the graph is as shown.

(1) What length of spring is shown by 1 small division (2 mm) on the vertical axis?
(2) What weight in kg is shown by 1 cm along the horizontal axis?
(3) Find what would be the length of the spring if the following weights were attached:
    (a) 2 kg, (b) 3·8 kg, (c) 5·6 kg, (d) 7·4 kg, (e) 11·2 kg.
(4) Find what weights would have to be attached to make the length
    (a) 15·5 cm, (b) 17·0 cm, (c) 21·1 cm, (d) 24·5 cm, (e) 26·75 cm.
(5) Find the increase in length for a change in weight from
    (a) 2 kg to 6 kg, (b) 6 kg to 10 kg.
(6) Find, from Question 5, the increase in length per kg attached.
(7) What is the unstretched length of the spring?

Example 3

A Knitting Conversion

Mrs Green has a cardigan pattern for a five-year-old which she wishes to use again for her daughter, who is now twelve. The longest row in the original pattern has 72 stitches, and this she decided to increase to 90 stitches. She drew the graph opposite to help her to make other increases in the same proportion. In this graph the original number of stitches has been measured along the vertical axis.

Exercises

(1) Complete the table she used to draw the graph:

Original Number of Stitches	0	12	56	72
New Number of Stitches	0			90

(2) What does 0·5 cm represent on each axis?
(3) Read from the graph the new number of stitches to replace these numbers on the old pattern:
    (a) Ribbing   8, 14, 16.
    (b) Sleeves   11, 15, 19, 23, 24, 28, 39, 42, 47, 50.
    (c) Fronts   20, 21, 25, 30, 37.
    (d) Back   27, 35, 43, 51, 55, 59, 62, 66, 69, 72.

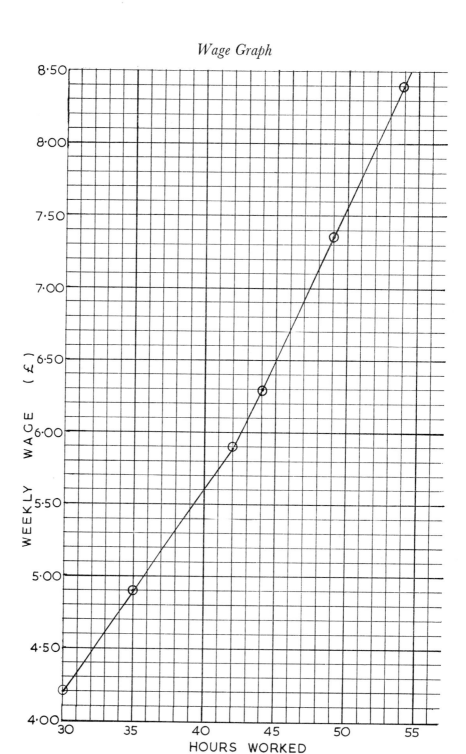

Example 4

Wage Graph (including Overtime)

This graph shows how the weekly wage of Joan, a sixteen-year-old girl, varies according to the number of hours worked.

Exercises

(1) How does the graph show the change from the ordinary rate to the overtime rate of pay?

(2) After how many hours per week does this change occur?

(3) Find Joan's wage for

   (*a*) 30 hours, (*b*) 42 hours, (*c*) 49 hours.

(4) Thus for 12 hours ordinary time the wage is . . .
   for 7 hours overtime the wage is . . .

(5) Now calculate:

   (*a*) the normal rate per hour,
   (*b*) the overtime rate per hour.

(6) Use the graph to find Joan's wage for a working week of

   (*a*) 32 hours, (*b*) 37 hours, (*c*) 44 hours, (*d*) 51 hours, (*e*) 54 hours.

(7) Find also how many hours per week Joan must work to earn

   (*a*) £4·76, (*b*) £5·32, (*c*) £6·72, (*d*) £7·56, (*e*) £7·98.

(8) If a youth of 17 years is paid 28p per hour for a 42-hour week, and 'time and a half' for overtime, use the graph to find his wage for

   (*a*) 33 hours, (*b*) 38 hours, (*c*) 43 hours, (*d*) 47 hours.

Celsius and Fahrenheit Temperatures

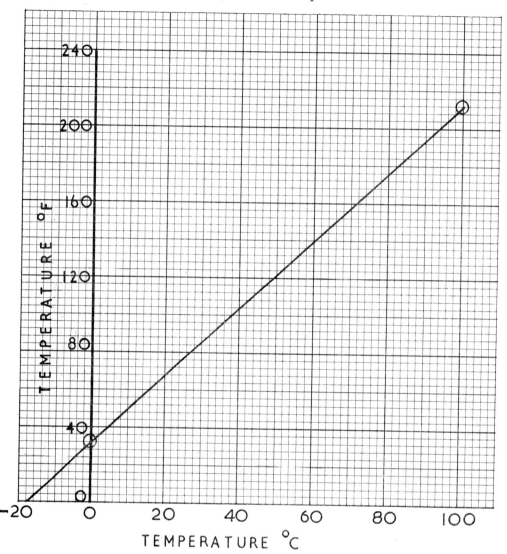

Example 5

Celsius and Fahrenheit Temperatures

Celsius and Fahrenheit thermometers are both used to measure temperatures, but on different scales. Our graph will help us to convert temperatures on one scale to temperatures on the other. The table for this graph contains only two pairs of values.

	Freezing-point (F.P.)	*Boiling-point* (B.P.)
Celsius temperature (°C)	0°	100°
Fahrenheit temperature (°F)	32°	212°

Exercises A

(1) State how many degrees are represented by one 2 mm division

   (*a*) on the *Fahrenheit* scale, (*b*) on the *Celsius* scale.

(2) Temperatures to the left of the vertical axis on the Celsius scale are marked — (minus). Why?

(3) Use the graph to convert the following temperatures

   (*a*) to degrees *Fahrenheit*:
   15°C, 22°C, 38°C, 45°C, 64°C, 72°C, 95°C, −10°C,

   (*b*) to degree *Celsius*:
   41°F, 68°F, 88°F, 107°F, 131°F, 156°F, 200°F, 24°F, 0°F.

(4) (*a*) Say what a change from 20°C to 40°C produces on the Fahrenheit scale.

   (*b*) Say what a change from 40°C to 80°C produces on the Fahrenheit scale.

   (*c*) From these results show that a rise of 1°C produces the same rise in degrees Fahrenheit in each case.

The equation of the line (see page 84) is: $F = 32 + \frac{9}{5}C$. In the following exercises select in each part the answer from the group of four which you consider to be correct.

Exercises B

(1) 25°C corresponds to: 45°F, 67°F, 77°F, 86·4°F.

(2) 99·5°F corresponds to: 35°C, 37·5°C, 34·6°C, 36·5°C.

(3) The gradient of the line is: 32, $\frac{3}{4}$, $\frac{5}{9}$, $\frac{9}{5}$.

(4) The equation of the line can be rewritten in the form:
$C = \frac{5}{9}(F-32)$, $C = \frac{5}{9}F - 32$, $32 + 9C = 5F$, $5C = 9F + 32$.

A Money Conversion—£'s, Francs and Deutschmarks

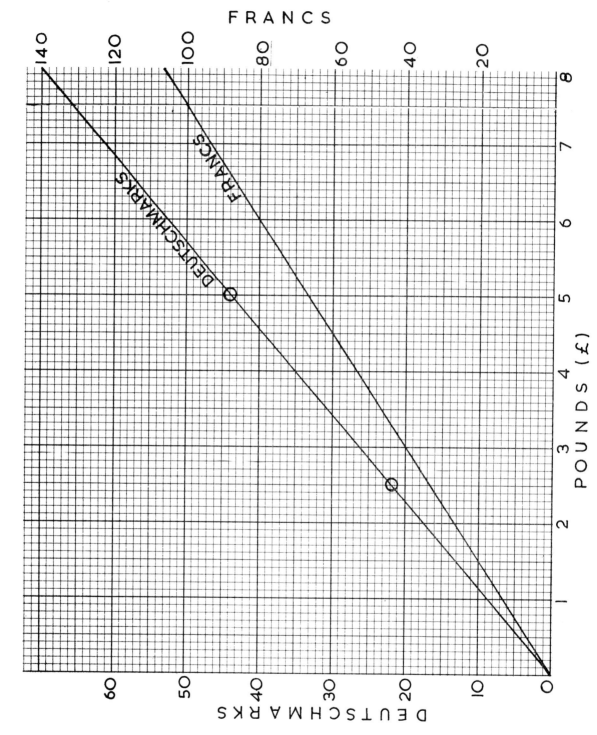

Example 6

A Money Conversion—£'s, Francs and Deutschmarks

The graphs on the opposite page show how two different vertical scales make it possible to convert

(a) £'s to Francs (fr.), (b) £'s to Deutschmarks (DM), (c) Francs to Deutschmarks.

Exercises

(1) Say how much each 2 mm division represents in

(a) Pence (p), (b) £'s, (c) Francs, (d) Deutschmarks.

(2) Find to the nearest unit the value

(a) in Francs, (b) in Deutschmarks

of the following:

£6, £2·10, £3·20, £3·75, £4·15, £5·40, £6·40, £7·15.

(3) Convert to £'s, to the nearest 5p:

(a) fr.:   20,   32,   48,   70,   89,   102.
(b) DM: 7,   19,   23,   34,   45,   62.

(4) Convert direct to the nearest Deutschmark:

Fr.:   12,   41,   56,   68,   83,   94.

(5) Convert direct to the nearest Franc:

DM: 21,   30,   39,   49,   58,   66.

(6) By selecting two suitable pairs of corresponding values from the results of Exercise 5, express

(a) 1 DM in Fr., (b) 1 fr. in DM.

EXERCISES ON DRAWING STRAIGHT LINE GRAPHS

(1) (a) A boy is paid $22\frac{1}{2}$p per hour for a 44-hour week. Draw a graph to show his wage for any number of hours from 20 to 44, calculating 4 values for your table.

(b) Use the graph to read off his wage for the following number of hours:

23,   25,   30,   35,   38,   43

and for what numbers of hours his wage would be:

£4·95,   £6·30,   £7·65,   £8·10,   £8·77$\frac{1}{2}$,   £9·45.

(2) (a) A traveller using his own car is given expenses of 3p per kilometre. Draw a graph showing his expenses (in £'s) for any number of kilometres up to 400.

(b) Read off his expenses for the following distances:

56,   92,   144,   224,   284,   360,   396

(c) and for what distances his expenses would be:

£3·90,   £5·10,   £6·84,   £10·08,   £11·61,   £13·53,   £14·10,   £15·27.

(3) Mr Smith is paid only commission on his sales at a rate of $4\frac{1}{2}$ per cent, whereas Mr Jones is given a basic weekly wage of £11·50 as well as commission on his sales at the rate of 2 per cent. Draw two graphs on the same graph sheet to show their weekly pay on sales up to £800. From the graph find how much Mr Smith must sell in a week to earn more than Mr Jones.

(4) In a mathematics examination the pupils' marks ranged from 12 to 84. The teacher decided to alter them so that the range was from 30 to 90.

Draw a graph showing the relation between the old mark (horizontal axis) and the new mark. From your graph find the new marks corresponding to old marks of:

27, 36, 42, 58, 63, 69, 71, 75, 78, 81.

By extending your graph until it meets the vertical axis and by finding the gradient of your graph, write down the equation of the graph connecting the new mark y and the old mark x.

(5) If 8 kilometres (km) = 5 miles, draw a graph for converting to km any number of miles up to 100, and vice versa. Use your graph

(a) to convert to km

25, 40, $47\frac{1}{2}$, 68, 77, $82\frac{1}{2}$, 90, 98 miles,

(b) to convert to miles

24, 44, 76, 90, 105, 132, 141, 153 km.

(6) If £1 = 2·40 dollars, draw a graph to change amounts up to £10 into dollars.

From the graph

(a) change to dollars

60p, 1·45, £2·10, £3·50, £4·$37\frac{1}{2}$, £6·80, £8·75, £9·40.

(b) change to £'s

3·80, 5·30, 8·95, 10·80, 12·35, 14·70, 19·15, 23·50 dollars.

(7) Mr Baker's normal working week is of 45 hours, for which his rate of pay is 48p per hour. For overtime his rate is 'time and a half'. Draw a graph to show Mr Baker's wage for any number of hours per week from 30 to 60.
Use your graph to find as accurately as you can

(a) Mr Baker's wage for

33, 39, 44, 51, 54, 60 hours.

(b) for how many hours in a week Mr Baker worked to earn

£16·80, £19·20, £20·40, £23·76, £27·36, £28·80, £31·68.

APPENDIX TO CHAPTER VII

Example 1

The dependent variable is sometimes called y and the vertical axis the y axis. The independent variable is called x and the horizontal axis the x axis.

From the graph table you will see that

$$\frac{y}{x} = \frac{\text{discount}}{\text{M.P.}} = \frac{6\tfrac{1}{2}}{100} = \frac{13}{200}$$

i.e. the discount $= \dfrac{13}{200}$ of the M.P.

or $$y = \frac{13}{200}x$$

which is called the equation of the graph.

Of course, if the M.P. (x) is zero, the discount (y) is zero, and this is shown by the fact that the graph passes through the origin where the x and y values are both zero.

If the discount rate were 10p in the £, then the discount would be one tenth of the M.P. and the equation would be

$$y = \frac{1}{10}x$$

Example 2

From a natural length of 13·4 cm the spring increases by 1·3 cm for every kg attached. So if we attach

(a) 1 kg the length = 13·4 cm + 1·3 × 1 cm
(b) 2 kg the length = 13·4 cm + 1·3 × 2 cm
(c) 3 kg the length = 13·4 cm + 1·3 × 3 cm

Thus the length = 13·4 cm + 1·3 times the weight in kg
and $y = 13\cdot4 + 1\cdot3x$ is the equation of the graph.

Example 3

If N = new number of stitches,
S = original number of stitches,

then
$$\frac{N}{S} = \frac{90}{72} = \frac{10}{8} = \frac{5}{4}$$

$$N = \frac{5}{4}S$$

or $y = \frac{5}{4}x$, which is the equation of this knitting graph.

Example 4

Up to 42 hours per week you will have found that for Joan
$$\text{wage (in £)} = 0.14 \times \text{hours worked.}$$
$$y = 0.14x.$$
This is the equation of the first part of the graph.

Above 42 hours per week the wage begins at £5·88 and an overtime wage in £ is added which is 0·21 times the extra number of hours (called, say, X).

Then wage (in £) = 5·88 + 0·21 times the number of hours overtime.
$$y = 5.88 + 0.21 X$$
which is the equation of the second part of the graph.

Example 5

The equation of this graph is
$$y = 32 + \frac{9}{5}x$$

or
$$F = 32 + \frac{9}{5}C, \text{ where } F = \text{temp. in } ° \text{ Fahrenheit}$$
$$C = \text{temp. in } ° \text{ Celsius}$$

By substituting values of C in this 'equation' we can find the corresponding values of F, and so check our conversions from °C to °F.

For example, if the temperature is 38°C.—*i.e.*, C = 38,
$$F = 32 + \frac{9}{5} \times 38$$
$$= 32 + \frac{342}{5}$$
$$= 32 + 68\frac{2}{5}$$
$$\fallingdotseq 100.$$

So the temperature corresponding to 38°C is 100°F (approx.).

Check your other conversions in Question 3(*a*) on page 79 in this way.

Now use the 'equation' in another form,
$$C = \frac{5}{9}(F - 32),$$
to check your conversions in Question 3(*b*).

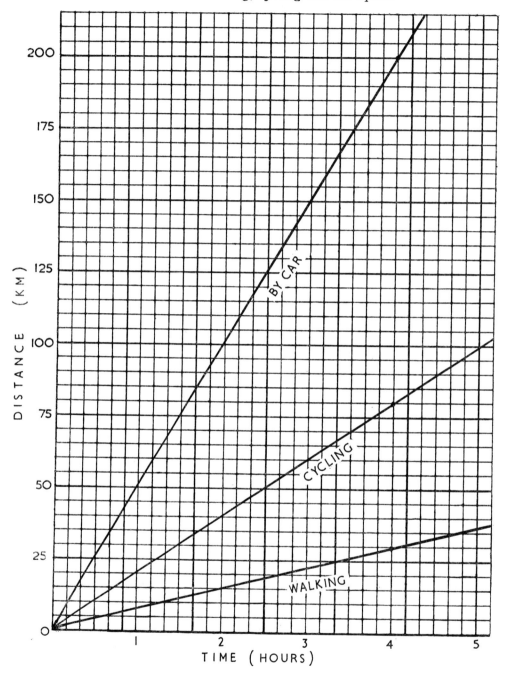

Distances at Walking, Cycling and Car Speeds

Chapter VIII

SPEED, TIME AND DISTANCE GRAPHS

If a vehicle moves at a steady speed the distance it travels will be in direct proportion to the time the vehicle is moving.

At 40 km/h, in 2 hours a car travels 80 km;

         in 4 hours the car travels 160 km.

So the graph of the distance travelled for various times will be a straight line.

(Although in actual fact the speed of a vehicle is continually varying during a journey, we treat our travel problems as though the vehicle were moving at a steady speed throughout. This is the average speed for the journey.)

In this chapter you should study
  (i) the effect of increased speed on the slope of a graph.
  (ii) the effect of a halt on a graph.
  (iii) the use of graphs for solving time and distance problems.

Example 1

Distances Travelled at Walking, Cycling and Car Speeds

The speeds chosen in drawing these graphs are reasonable average speeds for the three methods of travel.

Exercises

(1) Find from the graphs the steady speeds chosen for
    (i) walking, (ii) cycling, (iii) travel by car.
(2) How many minutes are there to each interval on the time scale?
(3) State the number of km to each interval on the distance scale.
(4) Copy and complete the following table, giving distances to the nearest kilometre for the times shown.

	3 hr	$2\frac{1}{2}$ hr	4 hr 10 min
Walking			
Cycling			
By car			

(5) Find in how many hours and minutes the car will travel
 (a) 27·5 km, (b) 116 km, (c) 296 km.
(6) What is the difference in distance travelled after $3\frac{1}{2}$ hours between
 (a) the walker and the cyclist, (b) the cyclist and the car.
(7) How is the slope of the graph affected by the speed of travel?

For algebra pupils

By car, the distance (in km) = 50 times the time in hours.

The equation of the graph for the car is $y = \ldots$

Similarly, the equation of the graph for cycling is $y = \ldots$, and the equation of the graph for walking is $y = \ldots$

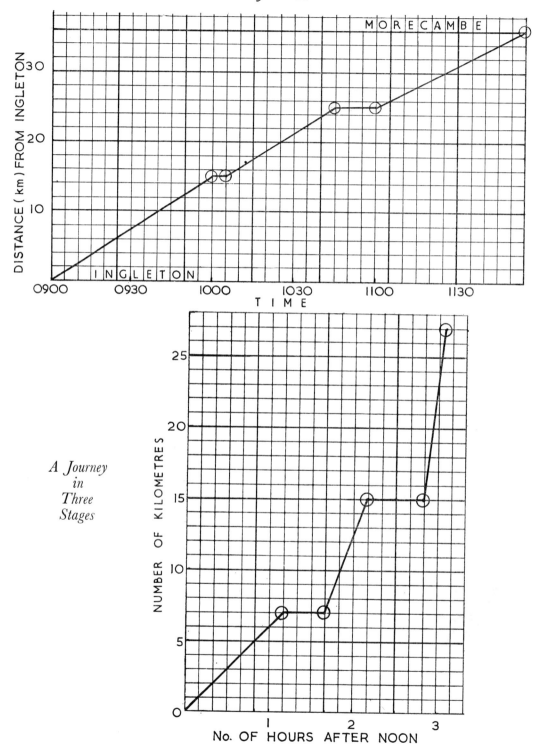

Example 2

A Cycle Run

The graph shows Kenneth's course on a cycle run from **Ingleton** to **Morecambe**. On the way Kenneth had two halts, the first to rest, the second because of a puncture. (Distances are measured in kilometres.)

Exercises

(1) On the horizontal axis say how many minutes are shown by 1 small division.

(2) At what speeds did Kenneth cycle on the 3 stages of his journey?

(3) Explain why the slope of the graph is the same for stages 1 and 2.

(4) How is a halt shown on the graph? Explain.

(5) Say how long Kenneth stopped

    (*a*) for a rest, (*b*) to mend a puncture.

(6) Say at what time Kenneth

    (*a*) left **Ingleton**,   (*b*) restarted after his rest,

    (*c*) had a puncture,   (*d*) reached **Morecambe**?

(7) Say how far from **Ingleton** is

    (*a*) the first halt, (*b*) the second halt, (*c*) **Morecambe**.

Example 3

A Journey in Three Stages

John Moore began a journey of 27 kilometres at noon and completed it in the following stages:

(1) He walked a km at a steady speed of u km/h, this stage taking x minutes. He then rested for r minutes.

(2) On the second stage of his journey he cycled b km at a steady speed of v km/h, taking y minutes on this stage. He rested again, this time for s minutes.

(3) Finally he travelled by car over the remaining c km at a steady speed of w km/h for z minutes.

Exercise

Use the graph on the opposite page to find the correct replacement values for $a, b, c, r, s, u, v, w, x, y, z$, showing your results clearly in a table.

A Time and Distance Problem

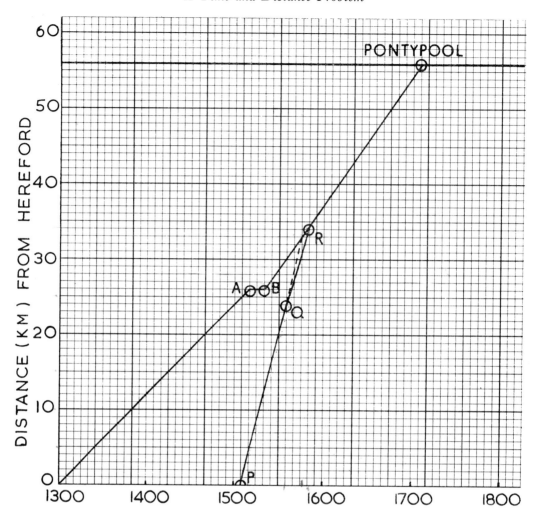

Example 4

A Time and Distance Problem

The graph OABR shows the distance from home of Joan, who set out from Hereford to cycle to her aunt's house in Pontypool 56 kilometres away. The graph PQR shows her father's journey. He followed her in his shooting brake to bring her home.

From the four suggested answers to each part of this example, select the one you decide is correct.

Exercises

(1) On the time axis 1 small division represents:
   (*a*) 10 min., (*b*) 6 min., (*c*) 5 min., (*d*) 4 min.
(2) Joan stopped at:
   (*a*) 1500 hours, (*b*) 1510 hours, (*c*) 1512 hours, (*d*) 1505 hours.
(3) When she stopped, Joan's distance from home was:
   (*a*) 24 km, (*b*) 30 km, (*c*) 25 km, (*d*) 26 km.
(4) Joan was stopped for:
   (*a*) 2 min., (*b*) 6 min., (*c*) 10 min., (*d*) 15 min.
(5) Joan's father left home at:
   (*a*) 1430 hours, (*b*) 1505 hours, (*c*) 1530 hours, (*d*) 1500 hours.
(6) He overtook Joan at:
   (*a*) 1550 hours, (*b*) 1545 hours, (*c*) 1540 hours, (*d*) 1555 hours.
(7) At this time Joan's distance from Pontypool was:
   (*a*) 20 km, (*b*) 23 km, (*c*) 34 km, (*d*) 22 km.
(8) Between O and A, Joan's average speed was:
   (*a*) 10 km/h, (*b*) 12 km/h, (*c*) 15 km/h, (*d*) 11 km/h.
(9) Between P and Q, her father's average speed was:
   (*a*) twice, (*b*) 3 times, (*c*) 5 times, (*d*) 4 times
   Joan's average speed between O and A.
(10) If he had maintained this speed from Q to R their meeting would have been earlier by approximately:
   (*a*) 2 min., (*b*) 10 min., (*c*) 4 min., (*d*) 8 min.
(11) Joan's average speed while cycling was approximately:
   (*a*) 10·5 km/h, (*b*) 13·5 km/h, (*c*) 12·75 km/h, (*d*) 15 km/h.
(12) The average speed of the car was approximately:
   (*a*) 45 km/h, (*b*) 35 km/h, (*c*) 38 km/h, (*d*) 41 km/h.

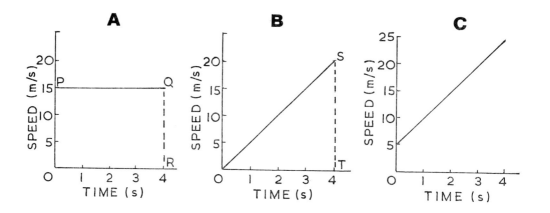
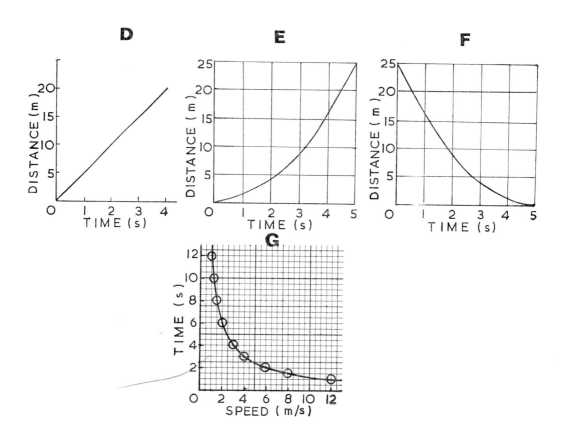

Example 5

The graphs on the opposite page illustrate:
- A. SPEED plotted against TIME for an object moving at a constant speed.
- B. SPEED plotted against TIME for an object starting from rest and moving with uniform acceleration.
- C. SPEED plotted against TIME for an object starting with an initial speed 5 m/s and moving with uniform acceleration.
- D. DISTANCE plotted against TIME for an object moving at a constant speed.
- E. DISTANCE plotted against TIME for an object moving with uniform acceleration. The graph is a parabola.
- F. DISTANCE plotted against TIME for an object moving with uniform deceleration. The graph is a parabola.
- G. TIME plotted against SPEED for a constant distance. The graph is a hyperbola.

Exercises

(1) In graph A
  (*a*) Write down the constant speed of the object.
  (*b*) Explain how the area of rectangle OPQR is a measure of the distance travelled at this speed.

(2) In graph B
  (*a*) Write down the uniform acceleration of the object.
  (*b*) What is its average speed over the 4 seconds?
  (*c*) Explain how the area of triangle OST is a measure of the distance travelled in 4 seconds at this average speed.

(3) In graph C
  (*a*) Write down the uniform acceleration of the object.
  (*b*) State how graphs B and C illustrate this same uniform acceleration.

(4) In graph D
  (*a*) Write down the distance travelled in each second and hence the constant speed of the object.
  (*b*) Explain how this graph shows that 'distance is proportional to time' if the speed remains constant.

(5) In graph E
  (*a*) Compile a table showing the distance travelled during each second from the 1st to the 5th second.
  (*b*) State the rate of uniform acceleration.

(6) In graph F
  (*a*) Compile a table as in question (5).
  (*b*) State the rate of uniform deceleration.

(7) In graph G
  (*a*) Write down the constant distance travelled.
  (*b*) State how the time varies with the speed.

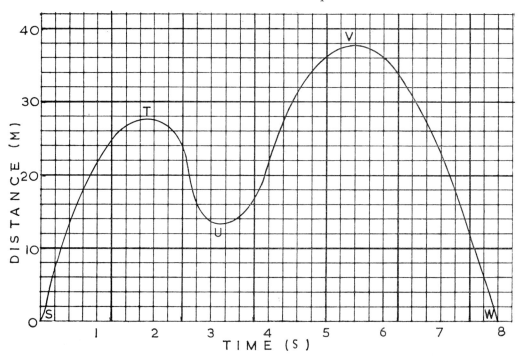

Example 6

A Distance–Time **Graph**

The graph shows the distance of a moving body from a starting point S at different times.

Exercises

(1) (*a*) Does the body return to its starting point?

   (*b*) If so, after how many seconds does it do so?

(2) Say between which points on the graph the distance of the body S from the starting point is

   (*a*) increasing, (*b*) decreasing.

(3) After how many seconds (approximately) does this distance first begin to decrease?

(4) Approximately how far from S is the body when its distance begins to increase again?

(5) State the maximum distance of the body from S and when this point is reached.

(6) Say how many times it passes through a point

   (*a*) approximately 27·7 metres from S,

   (*b*) 16 metres from S,

   (*c*) approximately 13·2 metres from S,

   (*d*) approximately 37·7 metres from S.

(7) Calculate the average speed of the body between the sixth and eighth seconds.

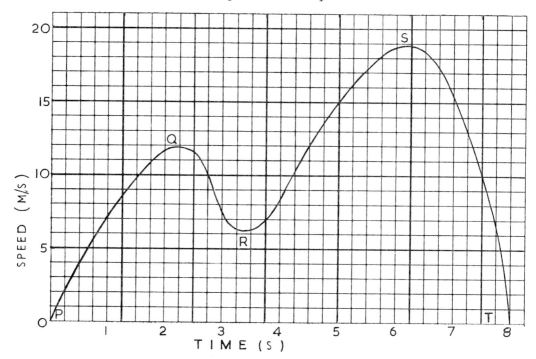
A Speed–Time Graph

Example 7

A Speed–Time Graph

The graph connects the speed of a body moving at varying speeds with the time during which it was observed. Points on the graph are marked P, Q, R, S, T to help in giving answers to the exercises.

Exercises

(1) State the initial velocity of the body.

(2) Say between which points the body was

    (*a*) accelerating, (*b*) decelerating.

(3) During the first 4 seconds give

    (*a*) the maximum velocity of the body and the time of its occurrence,

    (*b*) the minimum velocity of the body and the time of its occurrence.

(4) State the overall maximum velocity of the body.

(5) Give the average rate of acceleration

    (*a*) from the start to the end of 2 seconds,

    (*b*) from the 4th to the 6th second.

(6) Give the average rate of deceleration between 6·5 seconds and 8 seconds.

(7) Say whether the body returns to its starting point after 8 seconds.

(8) Did the body come to rest after 8 seconds?

(9) Q, R, S are 'turning points' of the graph. Explain in terms of acceleration and deceleration what happens to the body at these points.

A Motorist Approaches and Passes the Traffic Lights

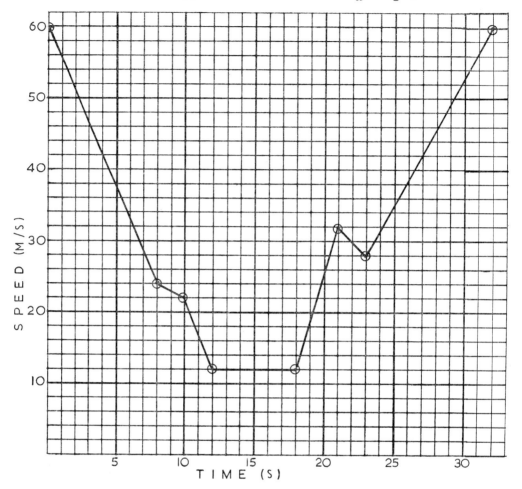

Example 8

A Motorist Approaches and Passes the Traffic Lights

A motorist travelling at a steady speed reduces speed when he sees the traffic lights at red ahead of him. He changes 'down', reduces speed still further and then proceeds at a steady speed until the lights are green. He then accelerates again, changes 'up' and continues to accelerate until he reaches his original steady speed six seconds after passing the traffic lights.

Use the graph on the opposite page to work the exercises.

Exercises

(1) At what steady speed was the motorist travelling when he first saw the traffic lights?
(2) At what speed was he travelling when he began to change 'down'?
(3) How many seconds did he take to execute the gear change?
(4) State in metres per second his rate of deceleration:
    (*a*) over the first 8 seconds,
    (*b*) over the 2 seconds following the gear change.
(5) At what steady reduced speed did he then travel until the lights changed to green?
(6) For how many seconds did he travel at this speed?
(7) Calculate in metres the distance the car travelled during this time.
(8) State his rates of acceleration, (*a*) before, and (*b*) after he changed 'up'.
(9) At what speed did he pass the traffic lights?

EXERCISES ON DRAWING TIME AND DISTANCE GRAPHS

(1) On the same 2 mm graph sheet draw graphs to show distances travelled for any time up to 6 hours at the following speeds:

(a) 10 km/h (walking), (b) 15 km/h (cycling), (c) 48 km/h (by car).

Use 24 mm to 1 hour on the time scale, and 20 mm to 30 km on the distance scale. From your graph read off (i) How far you would travel by each method in

(a) $1\frac{1}{2}$ hr., (b) 2 hr. 40 min., (c) 3 hr. 15 min., (d) 4 hr. 50 min.

(ii) How long you would take by each method to travel

(a) 10 km, (b) 18 km, (c) 45 km, (d) 50 km.

(2) A train travels 32 km in 24 minutes.
Draw a graph showing the distance travelled in any time up to 60 minutes if the train maintains this steady speed. Use your graph

(i) to find the distance it would travel in
5, 8, 17, 28, 36, 47, 51, 55 minutes,

(ii) to find the time it would take to travel
9, 15, 18, $28\frac{1}{2}$, $40\frac{1}{2}$, 54, $67\frac{1}{2}$, 76 km.

(3) Tom and David set out from home at 1100 hours on a cycle run at 10·5 km/h. After 2 hr. 40 min. cycling they rested for 20 minutes. Then they cycled home at 16 km/h. Draw a graph of their journey, and find at what time they arrived home.

(4) John leaves **Carlisle** at 0800 hours to walk at 7 km/h to **Penrith**, a distance of 29 km. His brother Martin follows at 1010 hours by cycle at 18 km/h. When and where will Martin overtake John?

(5) A car leaves **Exeter** at 1000 hours travelling at 40 km/h for **Ilfracombe**, 86·5 km away. A second car leaves **Ilfracombe** for **Exeter** at 1100 hours, travelling at 48 km/h. How far from **Exeter** and at what time will they pass each other?

(6) A motorist makes a journey of 128 km in two separate stages. The first is 48 km long, which the motorist covers at an average speed of 64 km/h. He rests for 15 minutes and then completes the second stage in 1 hour 15 min.

Draw a graph to illustrate his journey and find

(a) His total time for the journey.

(b) His average speed for the second stage.

(c) His average speed for the complete journey.

(d) His average speed if he did not rest.

(7) Mary Turnbull leaves school for home at 1550 hours. In the next 10 minutes she walks 1 km to the bus stop. She waits till 1605 hours for the bus which takes her 6 km to the stop 1 km from her home by 1620 hours. If she arrives home at 1635 hours draw a graph to show Mary's journey and find

   (a) the average speed of the bus,
   (b) Mary's walking speed,
       (i) from the school to the first bus stop,
       (ii) from the second bus stop to her home.

(8) Here is an extract from a railway time-table.

Distance (km)	Station	Arr.	Dep.
0	Bristol		0745
17·5	Bath	0802	0805
38·5	Chippenham	0823	0825
104	Didcot	0915	—
189	Paddington	1010	—

Draw a graph showing the 4 stages of the journey and find

   (a) which was the fastest part of the journey,
   (b) how long the train took to cover the first half of the journey.

(9) Here is an extract from a transatlantic air time-table showing the times of 2 aeroplanes, one from **London** to **Detroit**, one from **Detroit** to **London**. (Times have been corrected to Greenwich Mean Time on the 24-hour-clock system.)

Distance	Time (G.M.T.)				Time (G.M.T.)
0 km	0900	Dep.	London Airport	Arr.	2300
560	1010	Arr.	Glasgow (Prestwick)	Dep.	2150
	1055	Dep.		Arr.	2110
5440	1720	Arr.	Montreal	Dep.	1530
	1820	Dep.		Arr.	1430
6300	1950	Arr.	Detroit	Dep.	1300

Draw graphs of these journeys, and find when and where the planes pass each other.

(10) The distance from **Canterbury** to **Dover** is 24 km. Mr Smith sets out from **Canterbury** for **Dover** at noon, walking at 6 km/h to meet his friend Mr Brown, who leaves **Dover** at 1240 hours, walking towards **Canterbury** at 7 km/h. If each rests for 10 minutes after every hour of walking when will they meet, and how far from **Canterbury** will this be?
(Scale: for distance 2 cm = 5 km, for time 6 cm = 1 hour.)

Chapter IX

TIME SERIES AND TREND GRAPHS

Line graphs which show variations in quantities with the passage of time are often called 'Time Series Graphs'. These quantities range from monthly figures of unemployment or of exports and imports, most useful to government departments, to weekly sales figures of great value to business and manufacturing executives, and even to hourly prices on the Stock Exchange which are important to stockbrokers and investors.

The up and down movement in the graph helps us to see the following three types of variation:

(a) Short term or seasonal movements, such as the rise in unemployment figures in outdoor occupations during winter months.

(b) A long term or basic trend which might show an overall rising or falling tendency despite short term fluctuations.

(c) Random fluctuations which cut across other general movements and which may be caused by such events as strikes or changes in the international situation.

Short term movements may be useful to managements who may wish, for example, to take effective action to arrest a fall in sales, but as a basis for forecasting they are hazardous because much depends on unforeseen events. If, however, there is a well defined trend over a long period, it may well be possible to forecast the not-too-distant future with some certainty.

Most time series graphs illustrate a combination of types (a) and (b) and we shall consider here two simple methods of fitting a trend line on graphs of this type.

Method 1—A Freehand Curve. This is drawn between the points on the graph to leave roughly the same number of points on either side of the curve and with these points approximately evenly spaced in pairs from the curve. This method demands a fair degree of skill and judgement.

Method 2—Moving Averages. If the monthly sales of a firm were given for a year we would calculate the average of the sales for, say, three months at a time, thus:

(a) Jan., Feb., Mar., (b) Feb., Mar., Apr., (c) Mar., Apr., May, and so on.

The averages are plotted against the *middle month of the group* and the points joined by means of a series of straight lines.

Consider the following table showing a firm's monthly sales and a 3-monthly moving average:

1A *A Firm's Monthly Sales*

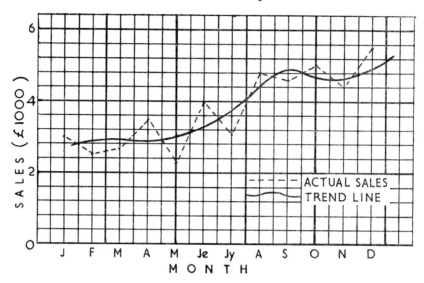

1B *A Firm's Monthly Sales*

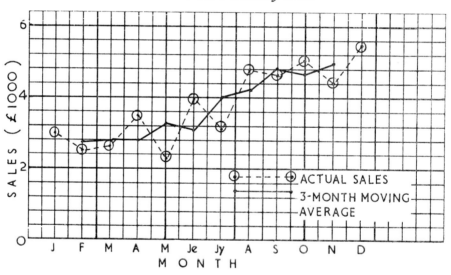

Month	Sales (£1000's)	3-month moving total (£1000's)	3-month moving average (£1000's) Trend
Jan.	3·0		
Feb.	2·5	8·1	2·7
Mar.	2·6	8·6	2·8
Apr.	3·5	8·4	2·8
May	2·3	9·8	3·3
June	4·0	9·4	3·1
July	3·1	11·9	4·0
Aug.	4·8	12·5	4·2
Sept.	4·6	14·4	4·8
Oct.	5·0	14·0	4·7
Nov.	4·4	14·9	5·0
Dec.	5·5		

The February figure in the third column is $\frac{3 \cdot 0 + 2 \cdot 5 + 2 \cdot 6}{3} = \frac{8 \cdot 1}{3} = 2 \cdot 7$. Graph 1A illustrates the original sales figures with a freehand curve to show the general trend.

Graph 1B replaces the freehand curve by the graph obtained from the 3-month moving average. It will be seen that this graph still shows fluctuations which are, however, much smaller than those of the original series.

A rough estimate of the figures for January and February of the following year could be obtained by projecting the trend line, but a more accurate method of further removing seasonal fluctuations is beyond the scope of this book.

Exercise

Construct a 5-month moving average for the sales given in the above table, copy the original sales graph and construct a trend line from your 5-month moving average.

Moving Annual Totals. When a time series is strongly seasonal the fluctuations in the corresponding graph may be smoothed out by working from an original amount for a year, and when the next month's figures are added to this yearly figure, those of the corresponding month of the previous year are deducted. The resulting graph shows annual totals from December to December, January to January, February to February, and so on.

Consider the sales of the Plantaida Fertilizer Company, shown quarterly for simplicity, in the following table:

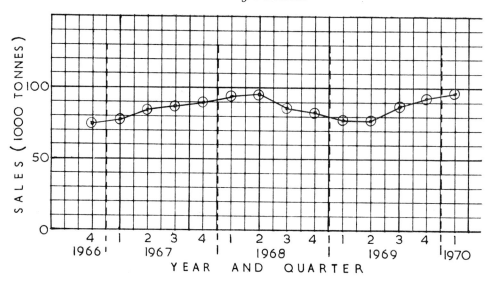
2 *Sales of Fertilizer*

'Plantaida Fertilizer Company'
Sales of fertilizer in thousand tonnes

Year and quarter		Sales	Moving annual total
1966	1	23	
	2	28	
	3	15	
	4	9	75
1967	1	27	79
	2	34	85
	3	18	88
	4	11	90
1968	1	31	94
	2	36	96
	3	10	88
	4	7	84
1969	1	26	79
	2	36	79
	3	19	88
	4	12	93
1970	1	30	97

In column 3 the figure 79 may be obtained by adding 27 (1967 1st quarter) to 75 and deducting 23 (1966 1st quarter), i.e., by adding $27 - 23 = 4$. This method of directly obtaining a trend type of graph is widely used. (See Graph 2.)

EXAMPLES ON CHAPTER IX

Note: The graph sheets with 0·5 cm divisions are sufficiently accurate for these graphs.

(1) The table shows the percentage of time lost through sickness in a factory over a period of 13 weeks. The totals of weeks 1 to 5, and 2 to 6, as well as the moving averages for these 5-week periods, have been entered opposite weeks 3 and 4 respectively. Complete the table and draw the original time series graph and a 5-week moving average graph on the same graph sheet.

Week	Percentage time lost	5-week total	5-week moving average
1	2·3		
2	3·5		
3	4·1	20·0	4·0
4	4·7	24·2	4·8
5	5·4	26·3	
6	6·5	29·3	
7	5·6	31·0	
8	7·1	29·9	
9	6·4	26·1	
10	4·3	20·6	
11	2·7	18·9	
12	3·1		
13	2·4		

(2) The table gives the daily number of school lunches served in a secondary school during the months of February and March. Calculate the total number of lunches for each week, placing the figures one below the other and inserting additional columns for 3-week totals and 3-week moving averages, respectively. Complete the table and construct the time series graph and a 3-week moving average graph on the same graph sheet.

	Week							
	1	2	3	4	5	6	7	8
Monday	136	132	120	115	102	100	102	104
Tuesday	128	122	124	110	114	106	103	105
Wednesday	116	118	110	112	108	103	100	99
Thursday	130	122	116	101	105	99	98	95
Friday	115	110	102	98	96	95	97	92

(3) The numbers of books borrowed from a library during successive months from January to December was:

765, 1013, 906, 675, 1063, 1002,
804, 1125, 1118, 1088, 1240, 1331.

Using a 3-month moving average, calculate the trend of the numbers of books borrowed. On the same graph sheet, construct the monthly time series graph and the 3-monthly moving average graph.

(4) The stocks of anthracite held by coal merchants in one area of Great Britain at the end of each quarter in years 1967 to 1969, inclusive, were (in thousands of tonnes):

Quarter	Year		
	1967	1968	1969
1st	16	28	23
2nd	20	39	36
3rd	21	40	38
4th	24	35	33

Plot these values as a time series and show on the same graph sheet a moving average which will eliminate a possible seasonal effect. (You should use a 4-quarterly moving average and the first value will be plotted midway between the values for the second and third quarter.)

(5) The quarterly sales of A.B. & Sons Ltd in £1000's are given in the table:

Year and quarter		Sales (£1000's)	Year and quarter		Sales (£1000's)
1966	1	58	1968	1	60
	2	35		2	36
	3	47		3	52
	4	77		4	84
1967	1	57	1969	1	62
	2	37		2	39
	3	49		3	53
	4	80		4	86

Plot these sales as a time series graph and show on the same graph sheet a 4-quarterly moving average which will eliminate a possible seasonal effect.

(6) Using the sales figures given in Example 5 and starting with the annual total of £217,000 for 1966, compile a table showing the moving annual total quarter by quarter. Construct a graph showing these moving annual totals.

(7) Using the figures for stocks of anthracite in Example 4, construct a graph of moving annual totals as in Example 6.

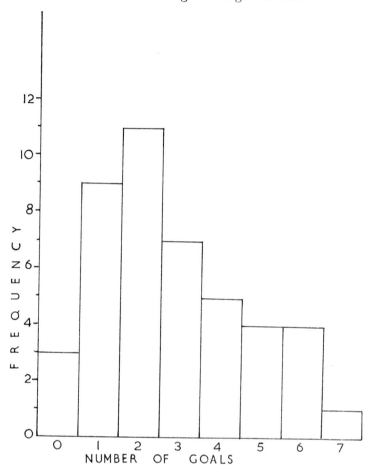

Goals Scored in English League Matches

Chapter X

STATISTICAL GRAPHS

Example 1

Goals Scored in English League Matches

The Histogram

This is a diagram used to illustrate a frequency distribution, whether, for example, it is symmetrical about a central value. Rectangles, based on the horizontal axis, are drawn with their areas proportional to the various frequencies. If the breadths of the rectangles are equal, their heights will be proportional to the frequencies.

The histogram shows the distribution of goals scored per match in the English Football Leagues (Divisions I to IV) on one recent Saturday.

Exercises

(1) Say in how many matches the following numbers of goals were scored:

   (*a*) 1, (*b*) 3, (*c*) 7.

(2) State the number of matches in which:

   (*a*) fewer than 4 goals were scored,

   (*b*) more than 2 goals were scored.

(3) Which is the '*modal*' number of goals scored.

(4) Find the total number of matches played.

(5) Calculate, correct to one decimal place, the average number of goals scored per match.

(6) Find in what percentage of the matches the '*modal*' number of goals was scored.

(7) In estimating the most likely number of goals in a particular match to be played the following week, say whether you would use your answer to Question 3 or Question 5, and give a reason for your choice.

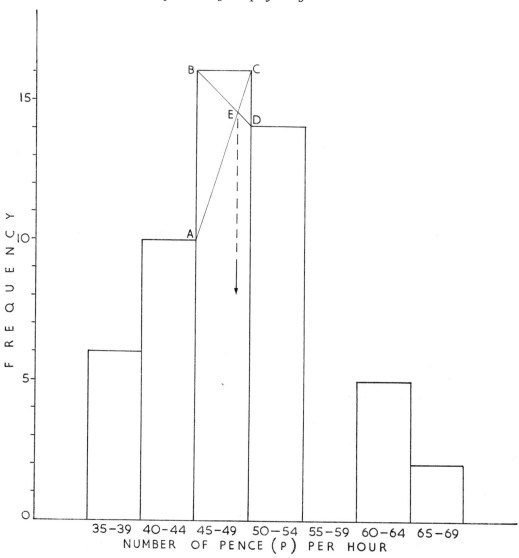
Hourly Rates of Employees of a Small Firm

Example 2

Hourly Rates of Employees of a Small Firm

This histogram illustrates how class intervals are used to show characteristics of a frequency distribution when the range of values is large.

Note: One column has been omitted from the histogram.

Exercises

(1) If the firm has 60 employees, how many of these should be shown in the missing column?

(2) Assuming that the average wage for each interval is the mid-interval value, find:

   (*a*) the total hourly pay of all employees.

   (*b*) their average hourly rate.

(3) Which class interval is the '*modal class*'?

(4) By joining A to C and B to D, their point of intersection E will give the approximate value of the *mode* from the 3 class intervals with the greatest frequencies. Estimate the mode.

(5) Find what percentage of the employees earn:

   (*a*) less than 45 pence per hour,

   (*b*) more than 52 pence per hour.

(6) State whether the variation in hourly rates is discrete or continuous.

Note: When it is intended to add a frequency polygon to a histogram one extra division is marked at each end of the horizontal axis, as shown in the example opposite.

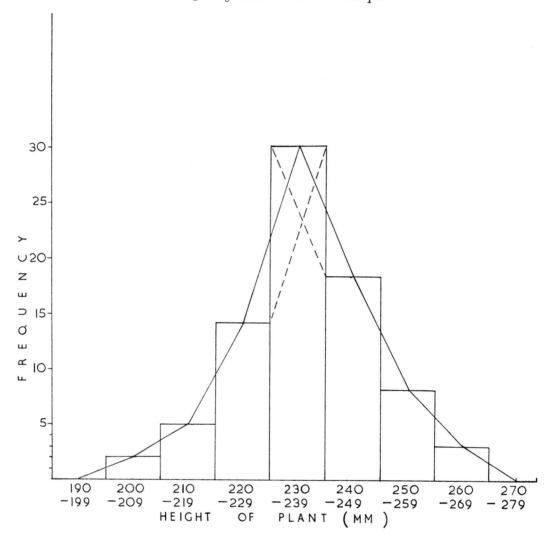

Example 3

Heights of Aster Plants in a Sample

The histogram and frequency polygon illustrate the frequency distribution of the heights of a sample of plants grown by a nurseryman. (The frequency polygon is formed by joining successive mid-points of the tops of the rectangles of the histogram, the end points being joined to the mid-points of extra divisions on the horizontal axis.)

Exercises

(1) State the 'modal' class of the heights.
(2) Find from the diagram the approximate value of the *'mode'* to the nearest mm.
(3) Check your answer to Question 2 by an arithmetical calculation.
(4) Explain briefly how the area 'under' the frequency polygon is the same as the area enclosed by the histogram.
(5) How many plants are in the sample?
(6) What percentage of the sample lies between 220 mm and 250 mm?
(7) Calculate the arithmetic mean of the heights of the plants.
(8) State whether the heights of the plants show discrete or continuous variation.

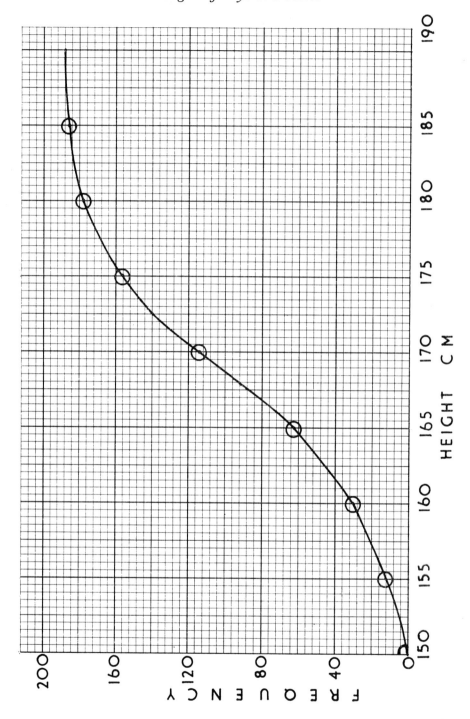
Heights of Boys in a School

Example 4

Heights of Boys in a School

The graph on the opposite page is an *'ogive'* or cumulative frequency curve. Use the graph to answer the following exercises.

Exercises

(1) How many boys are included in the diagram?
(2) What is the *median* height?
(3) How many boys were 166·5 cm or less?
(4) Calculate the percentage of boys over 172·5 cm.
(5) Find the *quartiles* and hence the *semi-interquartile range.**
(6) Using the gradient of the graph, estimate the 'modal' class.
(7) The following table shows some of the values from which the curve was drawn. Complete it and use it to find the *mean height* of the distribution.

Height (cm)	Frequency	Cumulative frequency
146-150	2	2
151-155	10	12
156-160	18	30

* Also known as the QUARTILE DEVIATION

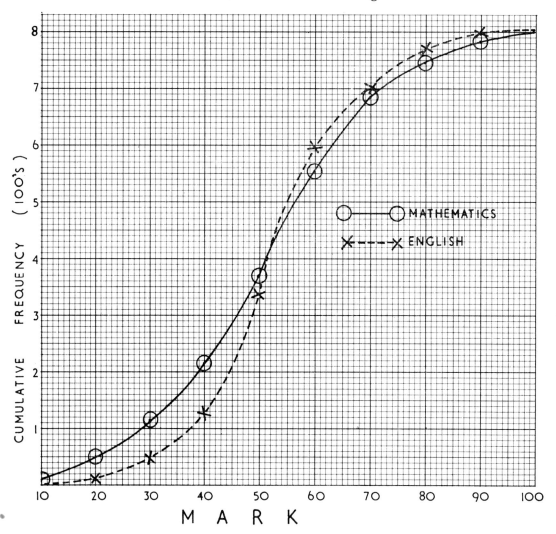

Example 5

CSE Results in Mathematics and English—a Comparison

The cumulative frequency curves on the opposite page give a comparison between results of one group of candidates in Mathematics and English in a recent CSE examination.

Exercises

(All answers should be given to the nearest whole number.)

(1) How many candidates were in the group?

(2) State the *median* mark in (*a*) Mathematics, (*b*) English.

(3) Find what percentage of the candidates scored

   (*a*) below 30%, (*b*) above 70%

   (i) in Mathematics, (ii) in English.

(4) Hence write down the percentage of candidates who scored between 30% and 70%

   (*a*) in Mathematics, (*b*) in English.

(5) If it is decided to allow 80% of the candidates to pass, state the pass mark (*a*) in Mathematics, (*b*) in English.

(6) Find the upper and lower 'quartiles' and use them to calculate the quartile deviations in each subject.

(7) Explain in what respect the quartile deviations illustrate a difference between the performance of the group in the two subjects, not shown by the 'medians'.

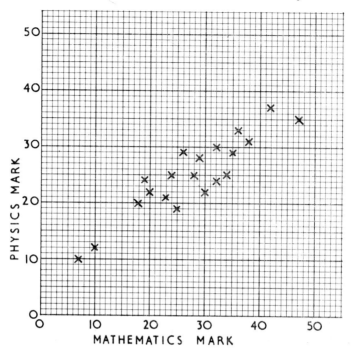
Correlation between Mathematics and Physics

Example 6

Correlation between Mathematics and Physics

The scatter diagram shows how, by plotting the Mathematics and Physics marks of twenty candidates (as x, y coordinates), a comparison of performances in these subjects may be made.

Exercises

(1) Copy the diagram on a graph sheet with small squares of 2 mm side.
(2) Draw in a line of best fit.
(3) Does the diagram show any positive correlation between the performances of the twenty candidates in the two subjects?
    Give a brief reason for your answer.
(4) Candidates scored marks in one of the subjects as shown below but were absent from the other examination:

Candidates	A	B	C	D
Mathematics mark	25	40	—	—
Physics marks	—	—	15	27

Use your diagram to estimate the missing marks.

O-level — CSE Comparison

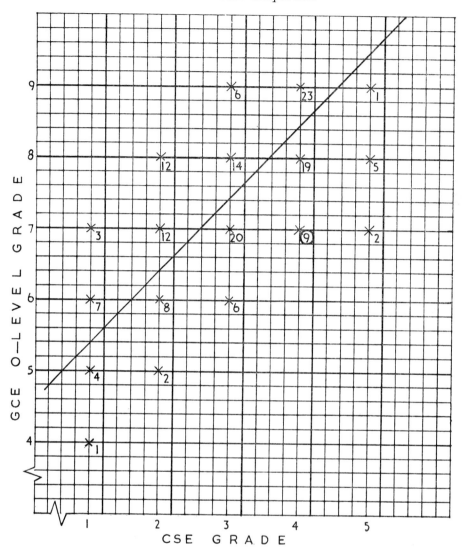

Example 7

Comparison of Grades in the CSE and GCE O-level Examinations

A number of candidates sat the same subject in the CSE and GCE examinations in the same year. The scatter diagram shows how many candidates obtained the two marked grades in these examinations, e.g., the circled number ⑨ indicates that 9 candidates obtained grade 4 in the CSE examination and also grade 7 in the GCE examination. A best fit line has been drawn to show a degree of correlation between the two sets of results.

Exercises

(1) Give two reasons why you consider the best fit line is fair and reasonably accurate.

(2) If the corresponding pass grades are reckoned as 5 and 6 in the GCE examination and 1 in the CSE examination, state briefly why the results as a whole show a positive correlation between these examinations.

(3) Calculate in each case what percentage of candidates failed in the GCE examination but gained passes at the following grades in the CSE examination:

(*a*) Grade 1, (*b*) Grade 2, (*c*) Grade 3.

(4) Find the average grade obtained by this group of candidates:

(*a*) in the CSE examination,

(*b*) in the GCE examination.

(5) Find what percentage of the group, who were regarded essentially as CSE candidates, passed the GCE examination.

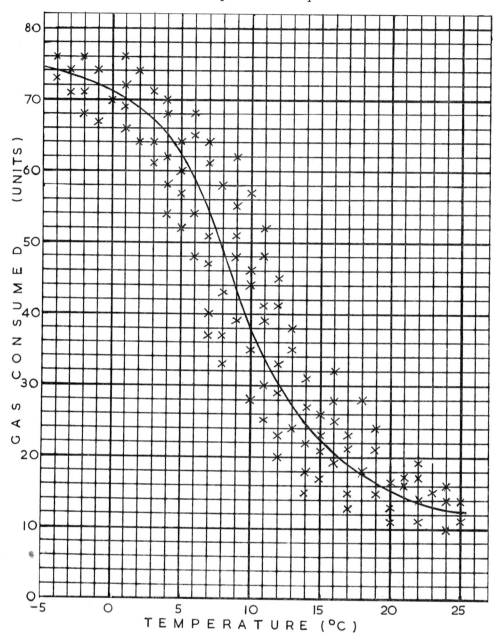

Gas Consumption and Temperature

Example 8

Gas Consumption and Temperature—a Scatter Diagram

A householder recorded over two years the number of units of gas he consumed each week, as well as the corresponding average temperatures. The gas was used for cooking, heating and lighting. The plotted diagram indicates how the amount of gas used is related to the temperature in degrees Celsius. The best fit line in this graph is curved.

Exercises

(1) Which type of correlation does the diagram show between gas used and temperature? Explain briefly.
(2) Between which range of 10 degrees C does the best fit line show a sharp drop in consumption?
(3) Give a reason why gas consumption levels out at the higher ranges of temperature.
(4) If the householder wished to reduce his gas account without suffering undue hardship, between which temperatures would you expect him to do so?

EXERCISES ON CHAPTER X

(1) The following are cricket scores made by members of a team in 60 individual innings:

```
10   1  14   0  11   0  25  27   0  34  17  16
 2  12   3   1  33   4  16  18   5  11  12  31
 0   1   0   9  13   8   5   6  10   7   5   7
 6   3   5  16   8  17   9   0   7   0   3   4
 7  21   5   6   2  24   7   5  22   2   6   8
```

(a) Using tally-marks make a grouped frequency distribution with an interval of 5 runs, for example, 0–4, 5–9, etc.

(b) State whether this distribution shows discrete or continuous variation.

(c) Construct the histogram and find the modal class.

(2) The following is the frequency distribution of weights of 100 women:

Weight (kg)	Frequency	Weight (kg)	Frequency
35–39	2	65–69	16
40–44	3	70–74	10
45–49	6	75–79	5
50–54	10	80–84	4
55–59	19	85–89	2
60–64	23		

Construct the histogram and the frequency polygon, and calculate the mean of distribution.

(3) The table shows the weekly profit of 120 shops.

Weekly Profit (£'s)	0–4	5–9	10–14	15–19	20–24	25–29	30–34	35–39	40–44	45–49
No. of Shops	2	5	10	17	28	26	15	7	6	4

(a) Construct the histogram and frequency polygon.

(b) If this sketch indicates how to obtain the approximate position of the mode from the class intervals with the three greatest frequencies, find its value from your histogram.

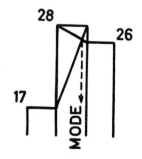

(4) (a) Draw a histogram to illustrate the number of times each of the vowels a, e, i, o, u occurs in the following verse of a well-known hymn:

'O Come, all ye faithful,
Joyful and triumphant,
O Come ye, O Come ye to Bethlehem;
Come and behold Him
Born the King of angels;
O Come, let us adore Him, Christ the Lord.'

(b) State which vowel is the mode.

(c) Justify by reference to the verse the statement that the relative frequencies are not characteristic of a normal sample of the English language.

(5) The table gives the number of deaths occurring over one year in each age group in a village of population 8600.

Age group	0–5	5–15	15–25	25–35	35–45	45–55	55–65	65–75
Deaths	8	3	3	4	6	11	14	15

Draw the histogram and note its U-shape. As the class interval 0–5 is only half the other class intervals, to make the area proportional to the frequency the height of this rectangle will have to be doubled.

(6) *Frequency.* To fix a piece-work rate a firm checked the number of articles completed per worker per day. The observations made over 12 days were as follows:

Worker
A	58	60	61	62	60	59	55	59	63	63	61	61
B	59	63	64	61	62	57	57	65	64	60	58	55
C	61	68	67	68	67	60	62	66	66	67	65	63
D	55	60	62	58	58	56	56	61	63	59	57	57
E	63	64	60	59	56	55	60	65	60	58	57	55
F	67	68	68	67	68	66	66	67	65	67	66	65
G	60	60	61	62	60	60	59	59	60	58	58	57
H	64	65	64	63	62	61	61	62	61	60	59	58

(a) Arrange these results in the form of a frequency table with headings: (i) *No. of Articles*, and (ii) *Frequency*.

(b) Draw the histogram and the frequency polygon.

(c) Which is the modal number of articles completed?

(d) If this modal number should give a basic pay rate of £2·70 per day, what should be paid per article?

(7) The wage rates per hour of 90 male employees in a firm are shown in the table:

Wage per hour (pence)	45	48	51	54	57	60	63
No. of men	5	12	20	24	16	9	4

(a) Find the mean wage rate per hour working from 54 pence as origin.

(b) Compile a cumulative frequency table and find the median rate.

(8) The following table gives the grouped scores of 200 children in an attainment test:

Score	Frequency	Score	Frequency
90–99	2	40–49	43
80–89	7	30–39	24
70–79	24	20–29	6
60–69	39	10–19	3
50–59	52		

(a) Using mid-interval scores of $94\frac{1}{2}$, $84\frac{1}{2}$, etc., calculate the mean of the group from an assumed mean of $54\frac{1}{2}$.

(b) Draw a cumulative frequency curve and use it to find the median, the quartiles and the quartile deviation.

(9) The basic weekly wages of the employees of a firm were:

8 earned between £17·01 and £18·00
10 earned between £18·01 and £19·00
19 earned between £19·01 and £20·00
13 earned between £20·01 and £21·00
9 earned between £21·01 and £22·00
4 earned between £22·01 and £23·00
2 earned between £23·01 and £24·00

(a) Construct an ogive to show the distribution.

(b) How many employees earned less than a basic £20 per week?

(c) What percentage of the employees earned over £21 per week (basic)?

(d) Find the median and the semi-interquartile range.

(10) During the first 10 weeks of a school term, house points gained by members of one class were:

Week	1	2	3	4	5	6	7	8	9	10
No. of points	4	8	5	6	3	9	7	10	6	9

(a) Draw the ogive to show the cumulative total of points week by week.

(b) Find the average number of points scored per week (correct to 1 decimal place).

(c) State also the greatest and least deviation from the average.

(d) Estimate the cumulative total of points at the end of the 11th week. How accurate would you expect your answer to be?

(11) The following table gives the number of seedlings of different heights:

Height (mm)	121–130	131–140	141–150	151–160	161–170	171–180	181–190
No. of seedlings	2	5	17	30	26	14	6

(a) Draw the cumulative frequency graph of the distribution.

(b) On your graph mark lines to show the median and the quartiles and write down your results.

(c) Find the semi-interquartile range.

(12) Twelve pupils obtained the following marks in English and French:

English mark	41	52	70	39	35	67	56	59	48	82	77	58
French mark	32	40	53	22	27	56	41	41	39	66	60	38

(a) Construct a scatter diagram and draw in your estimate of the line of best fit.

(b) Give two reasonable conclusions from your diagram.

(c) Pupil A scored 61 in English and pupil B scored 35 in French but they were each absent from the examination in the other subject. From your diagram, estimate a mark in French for pupil A and in English for pupil B.

(13)

	Age in years						
Boy	12	13	14	15	16	17	18
A	96	91	88	82	77	75	70
B	105	98	96	87	85	84	81
C	93	85	81	80	79	77	77

The table shows the average golf scores of three boys noted each year by a club professional.

(a) Construct a scatter diagram.

(b) Draw in the line of best fit.

(c) State the type and degree of correlation, giving a reason for your answer.

(d) Explain why you think this relationship would be a normal one for the quantities involved.

ANSWERS

(Where the answers are obvious or require a graph they have been omitted)

Chapter I

Example 1 (p. 9)
1. 1 viewer. **2.** (a) 3, (b) 8, (c) 6. **3.** (a) 5 to 8, (b) 13 to 16, (c) 21 to 24. **4.** 1 pupil does not watch TV. **5.** 2/9. **6.** (a) 20, (b) 8.

Example 2 (p. 11)
1. 5. **2.** (a) 5, (b) 4, (c) 6. **3.** (a) 4, (b) $1\frac{1}{2}$. **4.** (a) 29, (b) 45, (c) 22. **5.** (a) Feb., (b) Apr., (c) Aug. **6.** (a) Jan., (b) May. **7.** Feb. and July.

Example 3 (p. 13)
1. (a) 50, (b) 10, (c) 30, (d) 70. **2.** 1000–1200 hours, (b) 0600–0800 hours, (c) 1600–1800 hours. **3.** (a) 950, (b) 860, (c) 600. **4.** (a) 1400–1600 hours, (b) 0800–1000 hours. **5.** (a) 1600–1800 hours, 1200–1400 hours, 0800–1000 hours. (b) These periods involve the rush-hour traffic. **6.** (a) 6130, (b) 380.

Example 4 (p. 15)
1. (a) 2p, (b) 7p, (c) 14p. **2.** (a) Water, (b) highways, (c) education. **3.** (a) 7, (b) 4, (c) 1.
4. (a) (i) £0·45, (ii) 45%; (b) (i) £0·14, (ii) 14%; (c) (i) £0·06, (ii) 6%.
5. (a) £27 360 000, (b) £273 600.

Chapter II

Example 1 (p. 18)
1. Sector angles (degrees):   27   45   54   63   81   90
Periods per week: 3   5   6   7   9   10.
2. Suitable answers give to Science and Mathematics the largest allocation, say 12 periods and 7 periods respectively.

Example 2 (p. 19)
1. Sector angles (degrees): 14, 22, 48, 60, 94, 128. **2.** (a) 26·1%, (b) 6·1%, (c) 3·9%.
3. (a) £524 000 000, (b) £1 397 000 000.

Example 3 (p. 19)
1. 36:25. **2.** £144. **3.** £8:£20. **4.** (a) Clothes and pleasure, (b) clothes and savings.
5. (a) Rent and rates, housekeeping, heating.

(b)	1970	1971
Rent and rates	£26	£31·97
Housekeeping	£40	£48·00
Heating	£6	£8·07

6. (a) 144 degrees, (b) 120 degrees. **7.** £24 per annum.

Exercises on Chapter II (p. 20)
Sector angles to the nearest 0·5 degrees.
1. 126, 75·5, 45, 81, 32·5. **2.** 153, 90, 36, 54, 27. **3.** 108, 54, 183·5, 14·5.
4. 145, 68, 75·5, 51·5, 20.

Chapter III

Example 1 (p. 23)

1. 1 cm = 500 m, 0·1 cm = 50 m. **2.** (*a*) 2500 m, (*b*) 4100, (*c*) 7350 m. **3.** (*a*) 6 cm, (*b*) 11 cm, (*c*) 15·2 cm. **4.** Everest in Tibet. **5.** Nevis in Great Britain. **6.** (*a*) Ararat and Blanc, (*b*) 300 m. **7.** Approx. 3½ times.

	Column height	Actual height
Everest	17·5 cm	8750 m
Ararat	10·2 cm	5100 m
Blanc	9·3 cm	4650 m
Nevis	2·8 cm	1400 m
McKinley	12·1 cm	6050 m
Aconcagua	14·0 cm	7000 m

Example 2 (p. 25)

1. 10. **2.** 1 cm = 1 absentee. **3.** (*a*) 2, (*b*) 8, (*c*) 5. **4.** (*a*) Monday p.m., (*b*) Wednesday a.m. and Friday p.m., (*c*) Thursday a.m. **5.** (*a*) Wednesday, (*b*) Thursday. **6.** Tuesday. **7.** (*a*) Thursday, (*b*) Tuesday.

Example 3 (p. 25)

1. The number of pupils present. **2.** 1 cm = 5 pupil attendances, 2 mm = 1 pupil attendance. **3.** (*a*) 38, (*b*) 32, (*c*) 35. **4.** (*a*) Monday p.m., (*b*) Wednesday a.m. and Friday p.m., (*c*) Thursday a.m. **5.** (*a*) Wednesday, (*b*) Thursday. **6.** 40.

Example 4 (p. 27)

1. 1 large division = 50 cm, 2 small divisions = 20 cm.

2.

	(*a*)	(*b*)	(*c*)	(*d*)	(*e*)	(*f*)
Height of column (cm)	6·5	10·5	17	8·4	15·6	6·25
Depth of water (cm)	130	210	340	168	312	125

3. (*a*) 330 cm, (*b*) 254 cm, (*c*) 104 cm. **4.** (*a*) 31 May, (*b*) 30 Sept., (*c*) 30 Nov. **5.** Wet. The water depth increased from 180 cm to 254 cm. **6.** June. **7.** (*a*) 507, (*b*) 125, (*c*) 240·5 each in millions of litres. **8.** (*a*) 195 million litres, (*b*) Sept., (*c*) Nov.

Example 5 (p. 29)

1. 1 cm = 10 marks. **2.** 78, 18, 70, 46. **3.** (*a*) Jones 3rd exam, (*b*) Smith 2nd exam, (*c*) Brown 2nd exam, (*d*) Scott 1st exam. **4.** Smith and Todd. **5.** Brown. **6.** Scott.

Example 6 (p. 29)

1. 86, 89, 32, 42. **2.** (*a*) Scott, Geography, (*b*) Jones, Science, (*c*) Brown, History, (*d*) Smith, Mathematics. **3.** (*a*) Smith and Scott, (*b*) Jones and Todd. **4.** Scott. **5.** Smith and Brown. **6.** Scott and Todd.

Example 7 (p. 31)

1. Zambesi, Storstromsbroen, Tay. **2.** Storstromsbroen and Tay. **3.** Golden Gate. **4.** (i) 1880 m, (ii) 2500 m, (iii) 3460 m.

Example 8 (p. 31)

1. (*a*) 12 km/h and 18 km/h, (*b*) 25 km/h, (*c*) 32 km/h. **2.** (*a*) 18 km/h, (*b*) 25 km/h and 32 km/h.

3.

	Distance	
	(*a*)	(*b*)
(i)	3·6 m	6·1 m
(ii)	9·5 m	19 m
(iii)	14·5 m	28 m

4. 25 km/h, 32 km/h.

Chapter IV

Example 1 (p. 37)

1. 5 degrees C. **3.** Feb. 18 to Feb. 21. **4.** (*a*) Feb. 14 to Feb. 20, (*b*) Feb. 21 to Feb. 24.
5. Feb. 16 to Feb. 22. **6.** No. **7.** Yes.

Example 2 (p. 37)

1. (*a*) 0700 hours and 1300 hours. (*b*) 1400 hours and 1900 hours. **2.** 1300 hours and 1400 hours.
3. (*a*) 4 degrees C, (*b*) 6 degrees C, (*c*) 0 degrees C. **4.** (*a*) 1300 hours and 1400 hours.
(*b*) Between 1100 hours and 1200 hours, and between 1600 hours and 1700 hours.
(*c*) 0800 hours and 1800 hours. **5.** (*a*) 5 degrees C, (*b*) 1130 hours. **7.** 10 minutes.

Example 3 (p. 38)

1. 1 cm = 2·5 kg. **3.** (*a*) 2·5 kg, (*b*) 10·6 kg. **4.** (*a*) From birth to age 3 months,
(*b*) from 15 months onwards. **5.** (*a*) 11·3 kg, (*b*) $13\frac{1}{2}$ months. **6.** 3.

Exercises on Chapter IV (p. 39)

2. 76 degrees C, 90 degrees C, 98 degrees C, 99·75 degrees C. **3.** No. In the time interval too great variations in height could occur. Over a much shorter time interval a rate of climb or drop might help.

Chapter V

Introduction

1. Age. **2.** Day of month. **3.** Cost of article. **4.** Time of day. **5.** Weights of letters. **6.** Speed.
7. Distance from Dover. **8.** Weights. **9.** Age in years. **10.** Age next birthday.

Examples 1A, 1B (p. 43)

1. (*a*) £18, (*b*) £48. **2.** (*a*) Tuesday, (*b*) Friday. **3.** (*a*) Friday, (*b*) Tuesday. **4.** Wednesday.
5. Thursday is pay day, (*b*) Money is becoming short before pay day.

Examples 2A, 2B (p. 45)

1. (*a*) 0, (*b*) 20. **2.** (*a*) 0·2 cm, (*b*) 1 cm. **3.** 5 times. **4.** (*a*) 2, (*b*) a horizontal line.
5. Wednesday. **6.** (*a*) Thursday, (*b*) Thursday. **7.** No.

Example 3 (p. 47)

1. 36°C. **2.** 5 cm = 1°C, 0·5 cm = 0·1°C. **3.** $\frac{1}{2}$ division (0·25 cm) = 1 hour. **4.** Every 4 hours.
5. (*a*) 0800 hours to 1600 hours, (*b*) 0000 hours to 1200 hours. **6.** (*a*) 1600 hours to 2400 hours,
(*b*) 1600 hours to 2000 hours. **7.** 1200 hours and 1600 hours on Wednesday. **8.** (*a*) 40°C,
(*b*) 38·3°C, (*c*) 37·4°C. **9.** 36·9°C. **10.** 0·5°C. **11.** (*a*) 39·2°C, (*b*) 38·1°C. **12.** 0800 hours on Wednesday.

Example 4 (p. 49)

1. (*a*) 2·5 cm, (*b*) 0·5 cm. **2.** (*a*) 1·5 cm, (*b*) 0·5 cm, (*c*) 5·5 cm. **3.** (*a*) 9, (*b*) 30, (*c*) 12·5,
(*d*) 9, (*e*) 10·7. **4.** (*a*) 1000 hours, (*b*) 1330 hours, (*c*) 1510 hours. **5.** Between 1000 hours and 1100 hours, and between 1700 hours and 1800 hours. **6.** (*a*) 54 km, (*b*) 24 km, (*c*) 84 km.
7. 28. **8.** 63 km. **9.** (*a*) $27\frac{1}{2}$, (*b*) $19\frac{1}{2}$, (*c*) $5\frac{1}{2}$. **10.** 1000 to 1100 and 1700 to 1800.

Example 5 (p. 51)

1. Moscow. **2.** (*a*) Winter, (*b*) Summer. **3.** Moscow.
4.

	Moscow	London
(*a*)	−10°C	4°C
(*b*)	20°C	18°C
(*c*)	11·5°C	14°C

5. (*a*) August, (*b*) February and December. **6.** (*a*) January, (*b*) May. **7.** Yes. **8.** Jan., Feb., March, Nov., Dec. **9.** 12°C. **10.** 3°C.

Example 6 (p. 53)

1. January and February; there is less traffic. **2.** (*a*) Jan., Feb., Mar. and Sept.
(*b*) Apr., May, June, July, Oct., Nov., Dec. (*c*) August.
3. Casualties fell below 30 000 for Nov. and Dec. 1967, and in 1968 were lower until Oct. than in either of the two previous years. **4.** Severe weather kept many vehicles off the roads.
5. Greater density of holiday traffic.

6.

	March	June	Nov.
1966	29 600	34 500	33 500
1967	30 500	31 000	29 000
1968	27 500	30 000	30 600

Chapter VI

Example 1 (p. 57)

1. Time of day. **2.** 15 minutes. **3.** (*a*) 0900 hours and 1700 hours, (*b*) 1300 hours. The sun is low in the sky early and late and high in the sky around midday. **4.** (*a*) Between 0900 hours and 1000 hours and between 1600 and 1700, (*b*) between 1200 hours and 1400 hours.
5. (*a*) 15·8 m, (*b*) 15·6 m, (*c*) 22 m, (*d*) 9·3 m. **6.** (*a*) 1100 hours and 1500 hours, (*b*) 1300 hours, (*c*) 0940 hours and 1620 hours. **7.** British Standard Time is one hour ahead of G.M.T.

Example 2 (p. 59)

1. (*a*) 4 cm = 5 m, hence 8 mm = 1 m. (*b*) 4 cm = 5 km, hence 8 mm = 1 km. **2.** (*a*) Ranges from sea level, (*b*) for upper ranges of height. **3.** (*a*) 4·9 km, (*b*) 10·3 km. **4.** (*a*) 4 m, (*b*) 11·4 m.
5. (*a*) 4·25 km. **6.** 5 m to 9·4 m.

Example 3 (p. 59)

1. (*a*) 1 hour, (*b*) 1 mm. **2.** (*a*) Wed. 0500 hours, (*b*) Tue. 0400 hours. **3.** (*a*) 1300 to 1400 hours, (*b*) 0000 to 0100 hours. **4.** (*a*) 749 mm, (*b*) 757·5 mm, (*c*) 765 mm. **5.** (*a*) Mon. 0000 hours, Tue. 0200 and 0600 hours; (*b*) Tue. 1800 hours, Wed. 1120 hours; (*c*) Mon. 0800 hours, Tue. 1430 hours.

Example 4 (p. 61)

1. (*a*) From 5 years to 19 years = 14 years, (*b*) from 100 cm to 180 cm = 80 cm.
2. (*a*) 14 years, 2 years, (*b*) 80 cm, 10 cm. **3.** From 10 years 4 months to 13 years 8 months approximately. **4.** (*a*) 13 years, (*b*) 16 years. **5.** 10 years 4 months and 13 years 8 months.
6. (*a*) 124 cm, (*b*) 146 cm, (*c*) 165·5 cm, (*d*) 175 cm. **7.** (*a*) 122·5 cm, (*b*) 149 cm, (*c*) 160 cm, (*d*) 161·5 cm. **8.** (*a*) 8 years 11 months, (*b*) 14 years 6 months. **9.** (*a*) 11 years 5 months, (*b*) 16 years to 17 years.

Example 5 (p. 63)

1. (*a*) 2 mm = 15 minutes, (*b*) 2 mm = 0·1 metres. **2.** (*a*) 0300 hours, (*b*) 4·7 m. **3.** (*a*) 1530 hours, (*b*) 12 h. 30 min. **4.** (*a*) 4·2 m, (*b*) 2·7 m, (*c*) 1·5 m, (*d*) 3·3 m, (*e*) 1·7 m. **5.** (*a*) 0610, 1245, 1900 hours, (*b*) 0110, 0420, 1330, 1730 hours. **6.** (*a*) 0015 hours to 0515 hours and 1305 hours to 1815 hours.

Example 6 (p. 65)

1. (*a*) £359, (*b*) £711. **2.** (*a*) £95·50, (*b*) £236·60. **3.** (*a*) 14 years, (*b*) 23 years. **4.** 20 years.
5. (*a*) £103, (*b*) £50. **6.** 25 years. **7.** (*a*) £240, (*b*) £1400. **8.** (*a*) £1500, (*b*) £615.

Exercises on Scales (p. 66)

	Paper A 1 division equals	Paper B 1 division equals
(a)	5 pupils	5 pupils
(b)	2 m	2·5 m
(c)	£20	£20
(d)	10 km	—
(e)	1 degree	1 degree
(f)	5 kg	5 kg
(g)	5 eggs	5 eggs
(h)	30 accidents	30 accidents
(i)	5 days	5 days
(j)	£40	£35
(k)	2 cm	2·5 cm
(l)	80 litres	80 litres

Exercises on Chapter VI (p. 67)

1. (a) 47 m, (b) after 6·5 seconds, (c) 99 m, 4·5 seconds. **2.** (a) 18·5 cm², (b) 2·7 cm.
3. (a) From 11 years 3 months to 14 years 11 months, (b) 11 years and 15 years.
(c) No, the gradient of each graph is greater than 45°. **4.** (a) (i) £876, (ii) £486; (c) (i) £308,
(ii) £341; (e) (i) 21 years, (ii) 27 years. **5.** (a) (i) £565, (ii) £310; (b) (i) £391·50, (ii) £1087·50;
(d) (i) £1356, (ii) £1040, (iii) £496. **6.** (a) 13 to 14 g, (b) 30 g, (d) (i) 3·8 to 4 g, (ii) 8·4 g.
7. (a) 756 m², (b) AB = BC = 27·5 m. **8.** (a) 19·2 m, (b) 1·25 seconds or 2·75 seconds,
(c) 0·6 seconds to 3·4 seconds. **9.** (1) (a) 93 000, (b) 93 400, (c) 94 600; (2) (a) 1963, (b) 1969.
10. (a) 10% per annum, (b) (i) 2 years 9 months, (ii) £139, (iii) £177.

Chapter VII

Example 1 (p. 73)

1. 39p, 91p, £1·17, £1·76, £2·21, £2·96, £3·71, £4·36, £4·75, £5·10. **2.** £10, £17, £32, £40, £44, £54, £62, £70, £74, £76. **3.** £7·80, £8·84, £9·75. **4.** £12, £28·50, £50.

Example 2 (p. 75)

1. 0·5 m. **2.** 1 kg. **3.** (a) 16 cm, (b) 18 cm, (c) 20·4 cm, (d) 22·8 cm, (e) 28 cm. **4.** (a) 1·9 kg, (b) 3 kg, (c) 6·1 kg, (d) 8·6 kg, (e) 10·4 kg. **5.** (a) 5·4 cm, (b) 5·4 cm. **6.** 1·35 cm. **7.** 13·5 cm.

Example 3 (p. 75)

1. 15, 70. **2.** 4 stitches.
3. Ribbing 10, 18, 20.
 Sleeves 14, 19, 24, 29, 30, 35, 49, 53, 59, 63.
 Fronts 25, 26, 31, 38, 46.
 Back 34, 44, 54, 64, 69, 74, 78, 83, 86, 90.

Example 4 (p. 77)

1. It rises more steeply. **2.** 42 hours. **3.** (a) £4·20, (b) £5·88, (c) £7·35. **4.** £1·68, £1·47.
5. (a) 14p, (b) 21p. **6.** (a) £4·48, (b) £5·18, (c) £6·30, (d) £7·77, (e) £8·40. **7.** (a) 34 hours,
(b) 38 hours, (c) 46 hours, (d) 50 hours, (e) 52 hours. **8.** (a) £9·24, (b) £10·64, (c) £12·18,
(d) £13·86.

Example 5, Exercises A (p. 79)

1. (a) 4, (b) 2. **2.** They are below zero. **3.** (a) 59, 72, 100·5, 113, 147, 162, 203, 14. (b) 5, 20, 31, 42, 55, 69, 93, −4·5, −18. **4.** (a) 36, (b) 72, (c) rise of 1°C = rise of 1·8°F.

Exercises B
1. 77°. **2.** 37·5. **3.** $\frac{9}{5}$. **4.** $C = \frac{5}{9}(F - 32)$.

Example 6 (p. 81)
1. (a) 10, (b) 0·1, (c) 2, (d) 1. **2.** (a) Francs, 80, 28, 43, 50, 55, 72, 85, 95, (b) D Marks 53, 18, 28, 33, 37, 47, 56, 63. **3.** (a) £1·50, £2·40, £3·60, £5·30, £6·75, £7·70, (b) 80p, £2·20, £2·60, £3·85, £5·10, £7·05. **4.** 8, 27, 37, 45, 55, 63. **5.** 32, 46, 59, 74, 88, 100. **6.** (a) 1·5, (b) 0·66.

Exercises on Chapter VII (p. 81)
1. (b) £5·17½, £5·62½, £6·75, £7·87½, £8·55, £9·67½. (c) 22, 28, 34, 36, 39, 42. **2.** (b) £1·68, £2·76, £4·32, £6·72, £8·52, £10·80, £11·88. (c) 130, 170, 228, 336, 387, 451, 470, 509. **3.** £460. **4.** 43, 50, 55, 68, 72, 77, 79, 82, 85, 87. $y = \frac{5}{6}x + 20$. **5.** (a) 40, 64, 76, 109, 123, 132, 144, 157. (b) 15, 27·5, 47·5, 56·3, 65·6, 82·5, 88·1, 95·6. **6.** (a) $1·44, $3·48, $5·04, $8·40, $10·50, $16·32, $21·00, $22·56. (b) £1·58, £2·21, £3·73, £4·50, £5·14½, £6·12½, £7·98, £9·79. **7.** (a) £15·84, £18·72, £21·12, £25·92, £28·08, £32·40. (b) 35, 40, 42½, 48, 53, 55, 59.

Chapter VIII

Example 1 (p. 87)
1. (i) 7½ km/h, (ii) 20 km/h, (iii) 50 km/h. **2.** 10. **3.** 5 km.
4. Walking: 22½, 19, 31.
   Cycling: 60, 50, 83.
   By car: 150, 125, 208.
5. (a) 33 min, (b) 2 h 19 min, (c) 5 h 55 min. **6.** (a) 44 km, (b) 105 km. **7.** The greater the speed the steeper the slope.
For algebra pupils the equations are:
(a) By car $y = 50x$, (b) cycling $y = 20x$, (c) walking $y = 7\frac{1}{2}x$.

Example 2 (p. 91)
1. 5. **2.** 15 km/h, 15 km/h, 12 km/h. **3.** The speeds are equal. **4.** By a line parallel to the time axis. There is no increase in distance. **5.** (a) 5 minutes, (b) 15 minutes. **6.** (a) 0900 hours, (b) 1005 hours, (c) 1045 hours, (d) 1155 hours. **7.** (a) 15 km, (b) 25 km, (c) 36 km.

Example 3 (p. 91)

$a = 7$	$r = 30$	$u = 6$	$x = 70$
$b = 8$	$s = 40$	$v = 16$	$y = 30$
$c = 12$		$w = 48$	$z = 15$

Example 4 (p. 93)
1. 5 minutes. **2.** 1510 hours. **3.** 26 km. **4.** 10 minutes. **5.** 1505 hours. **6.** 1550 hours. **7.** 34 k.m **8.** 12 km/h. **9.** 4 times. **10.** 4 minutes. **11.** 13·5 km/h. **12.** 45 km/h.

Example 5 (p. 95)
1. 15 m/s; (b) the area of rectangle OPQR is the product of the time in seconds and the speed in metres per second. **2.** (a) 5 m/s; (b) 10 m/s; (c) the area of triangle OST = OS × ST/2 which is the time in seconds multiplied by the average speed in metres per second.
3. (a) 5 m/s; (b) each shows an increase in speed of 20 m/s in 5 seconds. **4.** (a) 5 metres, 5 metres per second; (b) the straight line graph illustrates that the two quantities compared are in proportion.
5. (a)

Second	1st	2nd	3rd	4th	5th
Distance	1 m	3 m	5 m	7 m	9 m

(b) 2 m/s/s.

6. (a)

Second	1st	2nd	3rd	4th	5th
Distance	9 m	7 m	5 m	3 m	1 m

(b) 2 m/s/s.

7. (*a*) 12 m, (*b*) inversely.

Example 6 (p. 97)

1. (*a*) Yes, (*b*) 8.   **2.** (*a*) Between S and T and between U and V; (*b*) Between T and U and between V and W.   **3.** 2 seconds.   **4.** 13·2 m.   **5.** 37·7 m after 5·5 seconds.   **6.** (*a*) 3 times, (*b*) 4 times, (*c*) 3 times (*d*) once.   **7.** 18 m/s.

Example 7 (p. 99)

1. Zero m/s.   **2.** (*a*) Between P and Q and between R and S; (*b*) Between Q and R and between S and T.   **3.** (*a*) 11·9 m/s after 2·25 seconds, (*b*) 6·2 m/s after 3·3 seconds.   **4.** 18·8 m/s. **5.** (*a*) 5·8 m/s/s, (*b*) 5·25 m/s/s.   **6.** 12·2 m/s/s.   **7.** Not known.   **8.** Yes.   **9.** The body speed is momentarily constant when acceleration stops and deceleration is about to begin (or vice versa).

Example 8 (p. 101)

1. 60 m/s.   **2.** 24 m/s.   **3.** Two.   **4.** (*a*) 4·5 m/s/s, (*b*) 5 m/s/s.   **5.** 12 m/s.   **6.** Six.   **7.** 72 m. **8.** (*a*) 6·7 m/s/s, (*b*) 3·6 m/s/s.   **9.** 39·2 m/s.

Exercises on Chapter VIII (p. 102)

		(*a*)	(*b*)	(*c*)	(*d*)
1.	(i) At 10 km/h	15 km	26·7 km	32·5 km	48·3 km
	At 15 km/h	22·5 km	40·0 km	48·8 km	72·5 km
	At 48 km/h	72·0 km	128·0 km	156·0 km	232·0 km
		(*a*)	(*b*)	(*c*)	(*d*)
	(ii) At 10 km/h	1 h	1 h 48 min	4½ h	5 h
	At 15 km/h	40 min	1 h 12 min	3 h	3 h 20 min
	At 48 km/h	12½ min	22½ min	56¼ min	62½ min.

2. (i) Distance (km): 6·7, 10·7, 22·7, 37·3, 48, 62·7, 68, 73·3.
  (ii) Time (min): 6·75, 11·25, 13·5, 21·4, 30·4, 40·5, 50·6, 57·0.

3. 1545 hours.   **4.** At approx. 1133 hours, 25 km from Penrith.   **5.** 61 km at 1132 hours. **6.** (*a*) 2 h 15 min, (*b*) 64 km/h, (*c*) 57 km/h, (*d*) 64 km/h.   **7.** (*a*) 24 km/h, (*b*) 6 km/h, 4 km/h. **8.** (*a*) From Didcot to Paddington, (*b*) 1 h 23 min.   **9.** At 1620 hours, 760 km from MONTREAL.   **10.** 1427 hours, 12·7 km from Canterbury.

Chapter IX

Exercises on Chapter IX (p. 110)

2. Weekly totals: 625, 604, 572, 536, 525, 503, 500, 495.
3-week moving averages: 600, 571, 544, 521, 509, 499.
3. 3-monthly moving averages: 895, 865, 881, 913, 956, 977, 1016, 1110, 1149, 1220.
4. 4-quarterly moving averages: 20·3, 23·3, 28·0, 32·8, 35·5, 34·3. 33·5, 33·0, 32·5.
5. 4-quarterly moving averages: 54·3, 54·0, 54·5, 55·0, 55·8, 56·5, 56·3, 57·0, 58·0, 58·5, 59·3, 59·5, 60·0.
6. Moving annual totals (£1000's): 217, 216, 218, 220, 223, 226, 225, 228, 232, 234, 237, 238, 240.
7. Moving annual totals: 81, 93, 112, 131, 142, 137, 134, 132, 130.

Chapter X

Example 1 (p. 113)

1. (a) 9, (b) 7, (c) 1. **2.** (a) 30, (b) 21. **3.** Two. **4.** 44. **5.** 2·8. **6.** 25%. **7.** The answer to Question 3; the average is affected by extreme cases.

Example 2 (p. 115)

1. 7. **2.** (a) £29·65, (b) 49·4p. **3.** 45 to 49p per hour. **4.** 48p per hour. **5.** (a) 26·7%, (b) 35%. **6.** Discrete.

Example 3 (p. 117)

1. 230–239 mm. **2.** 235 mm. **3.** $(229 \cdot 5 + 11 \cdot 5/20 \times 10)$ mm.
4. The area of each triangle cut off in forming the frequency polygon is equal to the area of the triangle which replaces it. **5.** 80. **6.** 77·5%. **7.** 236·3 mm. **8.** Continuous.

Example 4 (p. 119)

1. 188. **2.** 168·3 cm. **3.** 76. **4.** 25·5%. **5.** $Q_1 = 162·6$ cm, $Q_3 = 172·6$ cm, semi-interquartile range = 5 cm. **6.** Mean height = 168·3 cm. **7.** 166–170 cm.

Example 5 (p. 121)

1. 800. **2.** (a) 51, (b) 51. **3.** (a) (i) 14%, (ii) 6%; (b) (i) 14·4%, (ii) 11·9%.
4. (a) 71·6%, (b) 83·1%. **5.** (a) 35, (b) 42.
6. Mathematics: $Q_1 = 39$, $Q_3 = 63$, quartile deviation = 12
   English: $Q_1 = 44·4$, $Q_3 = 60·2$, quartile deviation = 7·9.
7. The dispersion is wider in Mathematics than in English.

Example 6 (p. 123)

3. Yes; the parallelogram shape of the scatter diagram moving upwards from left to right indicates that low/high marks in Mathematics correspond to low/high marks in Physics.

	A	B	C	D
4. Mathematics mark:	25	40	17	31
Physics mark:	23	33	15	27

Example 7 (p. 125)

1. It runs parallel to two sides of the parallelogram shape of the diagram; the results of half of the candidates lie on either side of the line. **2.** In each exam higher/lower gradings correspond, while greater numbers of candidates obtained intermediate than top and bottom gradings.
3. (a) 20%, (b) 71%, (c) 87%. **4.** (a) Grade 3, (b) Grade 7·4. **5.** 18·2%.

Example 8 (p. 127)

1. Negative; low temperatures correspond to high gas consumption and vice versa.
2. 5°C to 15°C. **3.** Most gas is then used for cooking, and this amount varies very little.
4. Between 10°C and 15°C.

Exercises on Chapter X (p. 128)

1. (a) Frequencies: 18, 20, 8, 6, 3, 2, 3. (b) discrete, (c) 5–9. **2.** 61·90 kg. **3.** (b) 23·7.
4. (a) Frequencies:
a	e	i	o	u
7	17	6	16	4

(b) the mode is "e", (c) repetition of "O Come" has increased the frequency of occurrence of the letter "o". **6.** (a) From 55 by whole numbers to 68, with respective frequencies of 5, 3, 6, 8, 8, 14, 10, 7, 7, 5, 6, 5, 7, 5. (c) 60, (d) 4½p. **7.** (a) 53·6p, (b) 52p.
8. (a) 54·25, (b) median = 54·1, quartile deviation = 10·8. **9.** (b) 37, (c) 23%,

(d) 1 median = £19·80, semi-interquartile range = £1·05½. **10.** (b) 6·7, (c) 3·7 and 0·3, (d) within point. **11.** (b) Median = 159·2 mm, Q1 = 150·8 mm, Q3 = 168·6 mm, semi-interquartile range = 8·9 mm. **12.** (b) There is a positive correlation between the English and the French marks. Most pupils will score less in French than in English. (c) Pupil A French mark: 48, Pupil B English mark: 48. **13.** (c) Negative; with increase in age the average scores decrease. (d) With experience and physical maturity the boys ought to produce lower scores.